Maintenance and Mechanics

The McGraw-Hill *CONTROLLING PILOT ERROR* Series

Weather
Terry T. Lankford

Communications
Paul E. Illman

Automation
Vladimir Risukhin

Controlled Flight into Terrain (CFIT/CFTT)
Daryl R. Smith

Training and Instruction
David A. Frazier

Checklists and Compliance
Thomas P. Turner

Maintenance and Mechanics
Larry Reithmaier

Situational Awareness
Paul A. Craig

Fatigue
James C. Miller

Culture, Environment, and CRM
Tony Kern

Cover Photo Credits (clockwise from upper left): PhotoDisc; Corbis Images; from *Spin Management and Recovery* by Michael C. Love; PhotoDisc; PhotoDisc; PhotoDisc; image by Kelly Parr; © 2001 Mike Fizer, all rights reserved; *Piloting Basics Handbook* by Bjork, courtesy of McGraw-Hill; PhotoDisc.

CONTROLLING PILOT ERROR

Maintenance and Mechanics

Larry Reithmaier

McGraw-Hill
New York Chicago San Francisco Lisbon London Madrid
Mexico City Milan New Delhi San Juan Seoul
Singapore Sydney Toronto

McGraw-Hill

A Division of The McGraw·Hill Companies

Copyright © 2001 by The McGraw-Hill Companies, Inc. All rights reserved. Printed in the United States of America. Except as permitted under the United States Copyright Act of 1976, no part of this publication may be reproduced or distributed in any form or by any means, or stored in a data base or retrieval system, without the prior written permission of the publisher.

1 2 3 4 5 6 7 8 9 0   DOC/DOC   0 7 6 5 4 3 2 1

ISBN 0-07-137319-5

The sponsoring editor for this book was Shelley Ingram Carr, the editing supervisor was David E. Fogarty, and the production supervisor was Sherri Souffrance. It was set in Garamond per the TAB3A design by McGraw-Hill Professional's Hightstown, N.J., composition unit.

Printed and bound by R. R. Donnelley & Sons Company.

 This book is printed on recycled, acid-free paper containing a minimum of 50% recycled de-inked fiber.

McGraw-Hill books are available at special quantity discounts to use as premiums and sales promotions, or for use in corporate training programs. For more information, please write to the Director of Special Sales, Professional Publishing, McGraw-Hill, Two Penn Plaza, New York, NY 10121-2298. Or contact your local bookstore.

Information contained in this work has been obtained by The McGraw-Hill Companies, Inc. ("McGraw-Hill") from sources believed to be reliable. However, neither McGraw-Hill nor its authors guarantee the accuracy or completeness of any information published herein, and neither McGraw-Hill nor its authors shall be responsible for any errors, omissions, or damages arising out of use of this information. This work is published with the understanding that McGraw-Hill and its authors are supplying information but are not attempting to render engineering or other professional services. If such services are required, the assistance of an appropriate professional should be sought.

Contents

Series Introduction *xiii*

Foreword *xxiii*

1 Introduction *1*
Federal Aviation Regulations (FAR) Maintenance Highlights *6*
FAA Aircraft Operating Certificates *9*
FAR Part 65—Certification of Airmen Other than Flight Crew *10*
Miscellaneous Definitions and Information *13*

2 Problems with MEL (Minimum Equipment List) *17*
Procedures Omitted for Disconnecting Constant-Speed Drive *20*
Cockpit Voice Recorder Inoperative and Deferred *22*
Inoperative Fuel Gauge per MEL *23*
Departure without Legally Required Documents *24*
Autopilot Altitude Hold Feature Inoperative *26*
MEL Item Not Repaired or Written Off in Log Book *28*
Delay due to Conflicting Interpretation of MEL by Maintenance *30*
Non-MEL Item Was Written Up as MEL Item *33*

Manufacturer Confirms Emergency Lights Not
Deferrable *36*
Gear Door Removed but Weight and Speed
Restrictions Not Listed *37*
Landing Gear Door Positioned Improperly *39*
Wingtip Light Inoperative and Deferred in
Noncompliance with MEL *41*
Window Cracked—Not Addressed by MEL *42*
MEL Operating Weight Restrictions Not Applied
Due to Software Failure *44*
Flight Dispatched in Noncompliance with Open,
Unanswered Log Report on Altimeter *45*
Inoperative Fuel Gauge and Incorrect Dripstick
Fuel Quantity *46*
Summary and Assessment: Problems with MEL *48*

3 Illegal and/or Unsafe Flight Due to Excessive Pressure for On-Time Departure *51*

MEL Not Complied with by Maintenance Due to
Pressure for On-Time Departure *54*
Instances of Deferred Maintenance a Dangerous
Policy *55*
Pressure for On-Time Performance Caused
Maintenance Problems *56*
Landing Gear Door Positioned Improperly *58*
Takeoff after Warning Horn and without
Maintenance Inspection *59*
Summary and Assessment: Illegal and/or Unsafe
Flight due to Excessive Pressure for On-Time
Departure *61*

4 Illegal or Unsafe Flight Due to Improper Maintenance Paperwork *65*

Flight Made without Properly Signed-Off
Maintenance Item *68*
Transport Flew Scheduled Passenger Flight
without Proper Maintenance Checks Being
Completed and Signed Off *69*

Contents *vii*

Plane Tagged as Airworthy by Maintenance When Beyond Annual Inspection *71*
Flight Took Off Again without Obtaining Rerelease from Dispatcher after Return for Pressurization Maintenance *72*
FAA Inspector Files Violations for Expired Registration and Lack of Weight and Balance Computation *74*
Owner-Pilot Did Some of His Own Maintenance, but Failed to Have Mechanic Sign Logbook *76*
Aircraft Airworthiness, Registration, and Maintenance Records Not in Order *78*
Flight Conducted with Unresolved Maintenance Items *80*
Captain Coerced into Accepting an Aircraft with Door Latch Problems *81*
Flight Crews Failed to Note That Logbook Lacked the Required Maintenance Release Sign-Off *83*
Aircraft Put Back in Operation with Repairs Completed but Not All Logbook Items Cleared *85*
Open Item in Maintenance Log *87*
Small Transport Grounded for Several Discrepancies, and FAA Held Last Pilot in Noncompliance *88*
Summary and Assessment: Illegal and/or Unsafe Flight Due to Improper Maintenance Paperwork *91*

5 Missed Checklist or Preflight Items *95*

Tail Damage Found Only After Several Legs Following Possible Tail Strike and Maintenance Inspection and Clearance *99*
Landing Gear Would Not Retract on Takeoff *100*
Engine Flameout and Emergency Landing Due to Reliance on Only a Verbal Assurance of Fuel Quantity *103*
Small Transport Descended below Assigned Altitude While on Autopilot *105*

viii Contents

    Flight Crew Raises Gear on a Gear-Down Ferry Flight   *106*
    Fire Bottles Not Hooked Up—Missed by Captain during Checklist   *109*
    Summary and Assessment: Missed Checklist or Preflight Items   *110*

6 Flight Crew Misunderstandings of Aircraft Systems Resulting in Illegal and/or Unsafe Flight   *113*

    Jet Engine Overtemperature during Start   *115*
    Transport Flew with Only One of Two Ignition Systems Working When Both Are Required   *116*
    Summary and Assessment: Flight Crew Misunderstanding of Aircraft Systems Resulting in Illegal and/or Unsafe Flight   *118*

7 Flight Crew Not Familiar with or Not Using Proper Abnormal or Emergency Procedures   *121*

    Fuel Imbalance Condition Misdiagnosed during Flight as Fuel Leak   *123*
    Transport Lands with One Gear Up Due to Disabled Uplock Pin Actuator   *126*
    Uncommanded Rudder Displacement, Crew Maintains Level Flight Using Standby Rudder on "B" Hydraulic System   *130*
    Summary and Assessment: Flight Crew Not Familiar with or Not Using Proper Abnormal or Emergency Procedures   *132*

8 Flight Crew Not Checking with Maintenance When Discrepancy Exists   *135*

    Subsequent Departure without Maintenance Sign-Off after In-Flight Engine Shutdown and Adding Oil to Gearbox That Overheated   *137*
    Aircraft with Skin Damage Flown without Proper Inspection   *139*

Contents **ix**

Electrical Problems Activate Warning Circuits during Aborted Takeoff and after Second Takeoff with No Maintenance Activity *141*

Pilot Performed Unauthorized Minor Maintenance Suggested by FAA Inspector *143*

Aircraft Departed with Open Maintenance Item in Logbook *145*

Departure without Clearance Regarding Maintenance Problem *146*

Captain Got Maintenance to Sign Off on Engine Check That Had Not Yet Been Done *148*

After Circuit Breaker Opened on a Previous Flight, Crew Failed to Make Logbook Entry and Get a Maintenance Write-Off before Next Departure *151*

Crew Made Fuel Transfer When Fuel Leak Noticed without Notifying Maintenance *153*

Pilot Advised of Violation Because Some Maintenance Items Had Not Been Signed Off by a Mechanic *154*

Aircraft Returns for Landing after Fire Warning Due to Fuel Leak after Installation of Fuel Nozzle *156*

Crew Proceeded with Takeoff without Consulting Maintenance after Aborting First Takeoff When They Noticed Reverser Pressure Light *158*

Commuter Transport Returned to Airport Instead of Diverting to Airport (Recommended by Maintenance) after Getting False-Warning on Master Panel *160*

Crew Starts Engines and Flies to Destination Even Though an Engine Exceeded Temperature Limits during First Start Attempt *162*

Pilot Departs Airport with Multiple Aircraft Faults Displayed on Warning Panels *164*

Crew Performed Takeoff with Fault Display Unit Showing No Information *165* x Contents

    Recurring Gear Extension Problems Not Entered in Logbook Because of Assumption of Ice Sticking to the Gear *166*
    Oxygen Service Door Was Left Open Causing Flight Control Difficulties and Emergency Landing *169*
    Freighter Dispatched with Left Navigation Light Inoperative in Darkness *171*
    Summary and Assessment: Flight Crews Not Checking with Maintenance When Discrepancy Exists *173*

9 Disagreements between Flight Crew and Maintenance and/or Management *175*

    Maintenance Inspection Waived to Avoid Delay *177*
    Pilot Questions Controller's Clearing for Takeoff in Instrument Conditions with Transport on 3-Mile Final *179*
    MEL Interpretation Dispute between Flight Crew and Maintenance *181*
    Captains of Two Transports Question Legality of Flying Aircraft with Electric Trim Inoperable *184*
    Captain's Complaint about Poor Company Maintenance *187*
    Transport Failed to Get Required 48-Hour Maintenance Check Prior to Departure *189*
    Owner and Pilot Performed Minor Maintenance *190*
    After Engine Start Fire, Cabin Attendant Deployed Emergency Exit without Pilot's Instruction *192*
    Flight Crew and Maintenance Disagree on Interpretation of MEL Requirements and Logbook Sign-Off *194*
    Captain Coerced into Accepting Aircraft with Door Latch Problems *196*

Captain Refused Aircraft after Maintenance
Examined Stabilizer, but Reconsidered after Chief
Pilot Examined and Accepted Aircraft   *198*
Flight Crew and Maintenance Disagree on
Whether Logbook Write-up Permits Flight   *200*
After Engine Exceeds Maximum Temperature
Flight Crew Deviates from Airline
Operational Procedure with Maintenance
Encouragement   *202*
Transport Dispatched in Questionable
Airworthiness Condition after Hail Damage
Repair Accomplished   *204*
Summary and Assessment: Disagreements
between Flight Crew and Maintenance or
Management   *206*

10 Summary and Final Discussion   *209*
Resolution of Error and Lessons Learned   *213*
Summary   *216*

Appendix Typical Turbine-Powered Airline Aircraft and Its Systems   *217*
The Aircraft   *219*
Design Features   *220*

Bibliography   *241*

Index   *243*

Series Introduction

The Human Condition

The Roman philosopher Cicero may have been the first to record the much-quoted phrase "to err is human." Since that time, for nearly 2000 years, the malady of human error has played out in triumph and tragedy. It has been the subject of countless doctoral dissertations, books, and, more recently, television documentaries such as "History's Greatest Military Blunders." Aviation is not exempt from this scrutiny, as evidenced by the excellent Learning Channel documentary "Blame the Pilot" or the NOVA special "Why Planes Crash," featuring John Nance. Indeed, error is so prevalent throughout history that our flaws have become associated with our very being, hence the phrase *the human condition*.

The Purpose of This Series

Simply stated, the Controlling Pilot Error series is to address the so-called human condition, improve performance in aviation, and, in so doing, save a few lives. It is not our intent to rehash the work of over a millennia of expert and amateur opinions but rather to *apply* some of the more important and insightful theoretical perspectives to the life and death arena of manned flight. To the best of my knowledge, no effort of this magnitude has ever been attempted in aviation, or anywhere else for that matter. What follows is an extraordinary combination of why, what, and how to avoid and control error in aviation.

Because most pilots are practical people at heart—many of whom like to spin a yarn over a cold lager—we will apply this wisdom to the daily flight environment, using a case study approach. The vast majority of the case studies you will read are taken directly from aviators who have made mistakes (or have been victimized by the mistakes of others) and survived to tell about it. Further to their credit, they have reported these events via the anonymous Aviation Safety Reporting System (ASRS), an outstanding program that provides a wealth of extremely useful and *usable* data to those who seek to make the skies a safer place.

A Brief Word about the ASRS

The ASRS was established in 1975 under a Memorandum of Agreement between the Federal Aviation Administration (FAA) and the National Aeronautics and Space Administration (NASA). According to the official ASRS web site, *http://asrs.arc.nasa.gov*

> The ASRS collects, analyzes, and responds to voluntarily submitted aviation safety incident reports in order to lessen the likelihood of aviation accidents. ASRS data are used to:
>
> - Identify deficiencies and discrepancies in the National Aviation System (NAS) so that these can be remedied by appropriate authorities.

- Support policy formulation and planning for, and improvements to, the NAS.
- Strengthen the foundation of aviation human factors safety research. This is particularly important since it is generally conceded *that over two-thirds of all aviation accidents and incidents have their roots in human performance errors* (emphasis added).

Certain types of analyses have already been done to the ASRS data to produce "data sets," or prepackaged groups of reports that have been screened "for the relevance to the topic description" (ASRS web site). These data sets serve as the foundation of our Controlling Pilot Error project. The data come *from* practitioners and are *for* practitioners.

The Great Debate

The title for this series was selected after much discussion and considerable debate. This is because many aviation professionals disagree about what should be done about the problem of pilot error. The debate is basically three sided. On one side are those who say we should seek any and all available means to *eliminate* human error from the cockpit. This effort takes on two forms. The first approach, backed by considerable capitalistic enthusiasm, is to automate human error out of the system. Literally billions of dollars are spent on so-called human-aiding technologies, high-tech systems such as the Ground Proximity Warning System (GPWS) and the Traffic Alert and Collision Avoidance System (TCAS). Although these systems have undoubtedly made the skies safer, some argue that they have made the pilot more complacent and dependent on the automation, creating an entirely new set of pilot errors. Already the automation enthusiasts are seeking robotic answers for this new challenge. Not surprisingly, many pilot trainers see the problem from a slightly different angle.

Another branch on the "eliminate error" side of the debate argues for higher training and education standards, more accountability, and better screening. This group (of which I count myself a member) argues that some industries (but not yet ours) simply don't make serious errors, or at least the errors are so infrequent that they are statistically nonexistent. This group asks, "How many errors should we allow those who handle nuclear weapons or highly dangerous viruses like Ebola or anthrax?" The group cites research on high-reliability organizations (HROs) and believes that aviation needs to be molded into the HRO mentality. (For more on high-reliability organizations, see *Culture, Environment, and CRM* in this series.) As you might expect, many status quo aviators don't warm quickly to these ideas for more education, training, and accountability—and point to their excellent safety records to say such efforts are not needed. They recommend a different approach, one where no one is really at fault.

On the far opposite side of the debate lie those who argue for "blameless cultures" and "error-tolerant systems." This group agrees with Cicero that "to err is human" and advocates "error-management," a concept that prepares pilots to recognize and "trap" error before it can build upon itself into a mishap chain of events. The group feels that training should be focused on primarily error mitigation rather than (or, in some cases, in addition to) error prevention.

Falling somewhere between these two extremes are two less-radical but still opposing ideas. The first approach is designed to prevent a recurring error. It goes something like this: "Pilot X did this or that and it led to a mishap, so don't do what Pilot X did." Regulators are particularly fond of this approach, and they attempt to regulate the last mishap out of future existence. These so-called rules written in blood provide the traditionalist with plenty of training materials and even come with ready-made case studies—the mishap that precipitated the rule.

Opponents to this "last mishap" philosophy argue for a more positive approach, one where we educate and train *toward* a complete set of known and valid competencies (positive behaviors) instead of seeking to eliminate negative behaviors. This group argues that the professional airmanship potential of the vast majority of our aviators is seldom approached—let alone realized. This was the subject of an earlier McGraw-Hill release, *Redefining Airmanship*.[1]

Who's Right? Who's Wrong? Who Cares?

It's not about *who's* right, but rather *what's* right. Taking the philosophy that there is value in all sides of a debate, the Controlling Pilot Error series is the first truly comprehensive approach to pilot error. By taking a unique "before-during-after" approach and using modern-era case studies, 10 authors—each an expert in the subject at hand—methodically attack the problem of pilot error from several angles. First, they focus on error prevention by taking a case study and showing how preemptive education and training, applied to planning and execution, could have avoided the error entirely. Second, the authors apply error management principles to the case study to show how a mistake could have been (or was) mitigated after it was made. Finally, the case study participants are treated to a thorough "debrief," where alternatives are discussed to prevent a reoccurrence of the error. By analyzing the conditions before, during, and after each case study, we hope to combine the best of all areas of the error-prevention debate.

A Word on Authors and Format

Topics and authors for this series were carefully analyzed and hand-picked. As mentioned earlier, the topics were taken from preculled data sets and selected for their relevance by NASA-Ames scientists. The authors were chosen for their interest and expertise in the given topic area. Some are experienced authors and researchers, but, more importantly, *all* are highly experienced in the aviation field about which they are writing. In a word, they are practitioners and have "been there and done that" as it relates to their particular topic.

In many cases, the authors have chosen to expand on the ASRS reports with case studies from a variety of sources, including their own experience. Although Controlling Pilot Error is designed as a comprehensive series, the reader should not expect complete uniformity of format or analytical approach. Each author has brought his own unique style and strengths to bear on the problem at hand. For this reason, each volume in the series can be used as a stand-alone reference or as a part of a complete library of common pilot error materials.

Although there are nearly as many ways to view pilot error as there are to make them, all authors were familiarized with what I personally believe should be the industry standard for the analysis of human error in aviation. The Human Factors Analysis and Classification System (HFACS) builds upon the groundbreaking and seminal work of James Reason to identify and organize human error into distinct and extremely useful subcategories. Scott Shappell and Doug Wiegmann completed the picture of error and error resistance by identifying common fail points in organizations and individuals. The following overview of this outstanding guide[2] to understanding pilot error is adapted from a United States Navy mishap investigation presentation.

Simply writing off aviation mishaps to "aircrew error" is a simplistic, if not naive, approach to mishap causation. After all, it is well established that mishaps cannot be attributed to a single cause, or in most instances, even a single individual. Rather, accidents are the end result of a myriad of latent and active failures, only the last of which are the unsafe acts of the aircrew.

As described by Reason,[3] active failures are the actions or inactions of operators that are believed to cause the accident. Traditionally referred to as "pilot error," they are the last "unsafe acts" committed by aircrew, often with immediate and tragic consequences. For example, forgetting to lower the landing gear before touch down or hotdogging through a box canyon will yield relatively immediate, and potentially grave, consequences.

In contrast, latent failures are errors committed by individuals within the supervisory chain of command that effect the tragic sequence of events characteristic of an accident. For example, it is not difficult to understand how tasking aviators at the expense of quality crew rest can lead to fatigue and ultimately errors (active failures) in the cockpit. Viewed from this perspective then, the unsafe acts of aircrew are the end result of a long chain of causes whose roots originate in other parts (often the upper echelons) of the organization. The problem is that these latent failures may lie dormant or undetected for hours, days, weeks, or longer until one day they bite the unsuspecting aircrew....

What makes [Reason's] "Swiss Cheese" model particularly useful in any investigation of pilot error is that it forces investigators to address latent failures within the causal sequence of events as well. For instance, latent failures such as fatigue, complacency, illness, and the loss of situational awareness all effect performance but can be overlooked by investigators with even the best of intentions. These particular latent failures are described within the context of the "Swiss Cheese" model as preconditions for unsafe acts. Likewise, unsafe supervisory practices can promote unsafe conditions within operators and ultimately unsafe acts will occur. Regardless, whenever a mishap does occur, the crew naturally bears a great deal of the responsibility and must be held accountable. However, in many instances, the latent failures at the supervisory level were equally, if not more, responsible for the mishap. In a sense, the crew was set up for failure....

But the "Swiss Cheese" model doesn't stop at the supervisory levels either; the organization itself can impact performance at all levels. For instance, in times of fiscal austerity funding is often cut, and as a result, training and flight time are curtailed. Supervisors are therefore left with tasking "non-proficient" aviators with sometimes-complex missions. Not surprisingly, causal factors such as task saturation and the loss of sit-

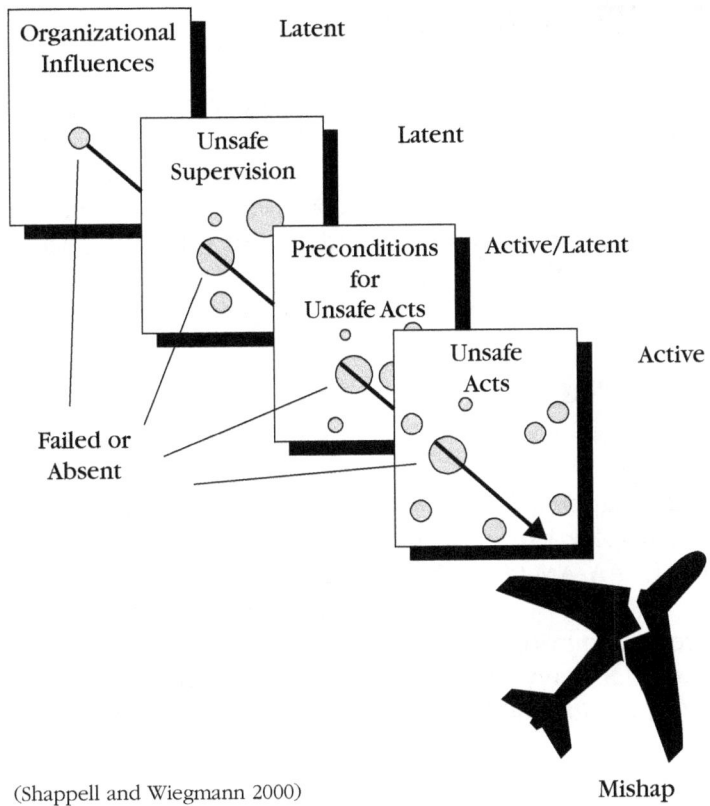

(Shappell and Wiegmann 2000)

uational awareness will begin to appear and consequently performance in the cockpit will suffer. As such, causal factors at all levels must be addressed if any mishap investigation and prevention system is going to work.[4]

The HFACS serves as a reference for error interpretation throughout this series, and we gratefully acknowledge the works of Drs. Reason, Shappell, and Wiegmann in this effort.

No Time to Lose

So let us begin a journey together toward greater knowledge, improved awareness, and safer skies. Pick up any volume in this series and begin the process of self-analysis that is required for significant personal or organizational change. The complexity of the aviation environment demands a foundation of solid airmanship and a healthy, positive approach to combating pilot error. We believe this series will help you on this quest.

References

1. Kern, Tony, *Redefining Airmanship,* McGraw-Hill, New York, 1997.

2. Shappell, S. A., and Wiegmann, D. A., *The Human Factors Analysis and Classification System—HFACS,* DOT/FAA/AM-00/7, February 2000.

3. Reason, J. T., *Human Error,* Cambridge University Press, Cambridge, England, 1990.

4. U.S. Navy, *A Human Error Approach to Accident Investigation,* OPNAV 3750.6R, Appendix O, 2000.

Tony Kern

Foreword

In 1918, the British journal *Lancet* published an article entitled "The Essential Characteristics of Successful and Unsuccessful Aviators: With Special Reference to Temperament." In it, two researchers reported on the findings of one of the very first studies of the psychological and physical predispositions of the pilot. They pointed out many truths that are taken as gospel by today's human factors experts, such as successful pilots "possess resolution, initiative, and presence of mind," and that unsuccessful pilots were "mentally sluggish, emotional, self-conscious and lacked a sense of humor." Experts were "good on the hunt," meaning that they were typically excellent horsemen with "good hands." Hidden toward the end of the essay, however, was an interesting item that, at first glance, seems to have been disproved over time. The researchers point out that based upon their observations, they firmly believe that pilots did not want or need to be mechanically inclined, because *if they understood what held the aircraft together and the aerodynamic principles that held the aircraft aloft, they might become too frightened to continue to fly at all.* While I feel we have overcome this fear, we may not have overcome the pilot's tendency to be less than mechanically inclined. This is a big mistake with lethal implications.

The purpose of this book is to remedy that situation in a user-friendly way. In case study after case study, you are about to be confronted with the mistakes that pilots, maintainance workers, and organizational supervisors are making in the field on a day-to-day basis. Taken directly from the files of the ASRS, and analyzed by the expert mind of Larry Reithmaier, these stories reflect in a very specific way where our greatest weaknesses lie. The author takes a realistic approach, pointing out the complexity of maintenance excellence balanced against organizational and individual communication cultures that can inhibit safe and effective operations.

In modern aviation, true knowledge does not necessarily mean having a full understanding of every subsystem of the aircraft. This would be impossible even for the most gifted of engineers, let alone the least gifted pilot. True knowledge, to quote Samuel Johnson, "is the ability to know where to *find* information when and where you need it." This book is one of those rare places where knowledge and application come together.

Inside these covers, you will gain insight and understanding on subjects such as the minimum equipment list, excessive pressure for on-time departure, improper maintenance paperwork, preflight checklist issues, improper pilot procedures, and how to manage disagreements between organizational supervision, maintenance personnel, and flight crew members. Best of all, you will be given actual scenarios to evaluate the actions of those who have "been there and done that." You will make your own decisions as to the correctness of their actions, and be more prepared to deal with these critical issues when they confront you personally.

The philosopher Elbert Hubbard once said, "The object of teaching was to prepare the student to get along well without the teacher." If this is true, then Larry Reithmaier has done his job. By the time you finish this book, you will have a far greater knowledge of what maintenance issues pose the greatest risk to the maintainer-pilot-machine team. Additionally, you will have a greater insight on the typical responses to these challenges, aided by Larry's expert analysis on what was done—and more importantly, what could have or should have been done to prevent the miscues from occurring in the first place.

The author of this book is superbly qualified to handle this difficult task. Larry Reithmaier is one of the top technical writers in aviation today. In addition to being an aircraft owner and pilot holding multiple certificates, Larry is a mechanical engineer with over 25 years of experience on the design and development of such aircraft as the F-86, F-100, F-101, F-4 and B-1 bomber. He has written a dozen books on aviation, including *Standard Aircraft Handbook for Mechanics and Technicians, Aviation/Space Dictionary,* and *Private Pilot's Guide.* He has been awarded multiple honors, including the top award presented by the Aviation/Space Writers Association.

So dig in and enjoy some of the finest analysis on the subject you will find on any bookshelf. You won't be disappointed.

Tony Kern

About the Author

Larry Reithmaier is a retired mechanical engineer who, while at Rockwell International, helped design and develop the F2H, F3, F86H, F100, F101, and F4 jet fighters, the B-1B bomber, and Apollo and Skylab spacecrafts. The author of several technical books on aviation, he also wrote the *Standard Aircraft Handbook for Mechanics and Technicians*, Sixth Edition; the *Aviation and Space Dictionary; Mach I and Beyond*; *Private Pilot's Guide*; and *Aircraft Repair Manual*.

1

Introduction

Aviation accidents due solely to maintenance deficiencies and oversight follow all other causes, with severe weather the number one cause. In general, the equipment we fly is safe and reliable. It's the way we operate it that gets us into trouble.

In the early years of aviation, it could be said that often the aircraft killed the pilot. Aircraft were intrinsically unforgiving and mechanically unsafe. However, the modern era has witnessed an ironic reversal of sorts. It now appears that the aircrew themselves are more dangerous than the aircraft. Human error has been implicated in more than two-thirds of all aviation accidents. Increased aircraft complexity and detailed and vigorously enforced regulations probably contribute to the human factors problem.

The design, testing, and subsequent certification of aircraft have evolved into an almost exact science. The flight crew and maintenance personnel have been presented with an intrinsically safe, if complex, aircraft.

Although the creation of aircraft has changed dramatically from the early years, the crew and maintenance personnel have the same characteristics. While engineers can assign numbers to their problems, the human factor specialists cannot be that exact. Improvement in the human factors element is the last major obstacle in the path to a zero accident rate.

Accidents, per se, are not the subject of this book. The National Transportation Safety Board (NTSB) investigates accidents that result in damage, injury, or death. They are reported because the law requires it. After all, we don't have that many accidents due to maintenance. We need to know about the incidents and hazards. Efforts to reduce incidents should also reduce accidents.

On December 1, 1974, a TWA B-727 was inbound from the northwest to land at Dulles International Airport in instrument meteorological conditions (IMC). The flight descended prematurely below the minimum safe instrument altitude, striking the slope at Mount Weather, Va. All 92 passengers and crew on board were killed.

Investigating the circumstances, the National Transportation Safety Board discovered that the flight crew misinterpreted information on the approach chart as well as ATC (air traffic control) instructions. The NTSB then discovered that another airline made a similar premature descent some six weeks earlier—somehow avoiding the same fatal error. *The earlier incident was reported within the company, but it was not disseminated to any other airlines for fear of enforcement action.*

This incident served as a catalyst to create an incident reporting system. Since one of the primary missions of the Federal Aviation Administration (FAA) is to promote aviation safety, the NTSB made an immediate recommendation for the FAA to create a reporting program designed to identify unsafe operating conditions. Thus the Aviation Safety Reporting Program [or System (ASRS)] was created.

The basic concept of the Federal Aviation Regulations (FAR) involves disciplinary action of some kind for violating one or more regulations, from suspension of a pilot certificate for a certain time period to monetary fines. If we expect to make progress in the human factors part of the safety equation, we have to encourage pilots to freely discuss their problems without fear of disciplinary action. It is a normal "human factor" to cover up or not admit our mistakes, transgressions, or oversights. If we do not grant anonymity, we certainly will not learn from human factors issues or problems.

In 1976, the FAA and the National Aeronautics and Space Administration (NASA) entered into a Memorandum of Agreement whereby NASA would handle the collection, analysis, and identification of safety reports. Although NASA designed and now administers the Aviation Safety Reporting System, the FAA provides the major funding for the ASRS to promote the continued use and operation of the system. NASA's ASRS is a voluntary, confidential incident reporting system that is designed primarily to provide information to the FAA and the aviation community to assist in reaching the goal of reducing, and ultimately eliminating, unsafe conditions in the National Aviation System (NAS). NASA's system also ensures the anonymity of the reporter.

Hazards or potential hazards are reported directly to NASA on their ASRS Reporting Form 277. In an effort to encourage the aviation community to report incidents, the FAA agrees that the entire form is not to be used as evidence to substantiate an alleged violation in an enforcement action.

The immunity provision is outlined in FAR part 91.25.

However, the immunity will not apply if the incident was deliberate, was criminal, or resulted in an accident. In addition, the reporter cannot have been involved in any enforcement action within the previous 5 years, and the incident must be reported to ASRS within 10 days of the event.

Because of the FAA's immunity provision and commitment to funding NASA's program, the ASRS now receives more than 34,000 reports per year. These reports are deidentified and held in strict confidence by NASA. For more than 20 years this arrangement has not been compromised.

Safety issues identified in the ASRS reports are discussed by FAA and NASA at a teleconference held twice a month. The issues can run the gamut from safety to security problems. In consultation with the ASRS program manager, NASA will forward to the FAA, aircraft manufacturers, airport representatives, and other aviation groups a special alert message that highlights the real or potential hazard.

Federal Aviation Regulations (FAR) Maintenance Highlights

To understand the real-life pilot reports in the following chapters, a brief summary of the Federal Aviation Regulations pertaining to maintenance is in order.

An overview of the FAR shows that the degree of regulation increases according to the extent of involvement of the general public. For example, the certification requirements for a single-reciprocating-engine, four-seat general aviation airplane are less stringent than for those of an air carrier, turbine-powered transport which is intended to carry 100 or more passengers for hire.

It is emphasized that the following FAR excerpts are not complete. They are included for information only and in sufficient detail to permit an understanding of the pilot reports in the following chapters.

Certification of pilots (FAR part 61)

This reference information is provided only in order to understand the pilot ratings of the ASRS incident reporters in the following chapters.

Pilot grades are as follows:

1. *Student pilots* have minimum certification requirements as well as maximum restrictions. They cannot carry passengers, among other restrictions.

2. *Recreational pilots* are restricted to carrying not more than one passenger in addition to many other restrictions.

3. *Private pilots* must pass required oral, written, and practical tests and can carry passengers not for hire.

4. *Commercial pilots* must pass increasingly stringent oral, written, and practical tests and can carry persons or property for hire among other privileges and restrictions.

5. *Airline transport pilots* must pass the most comprehensive oral, written, and practical tests. An airline transport certificate is normally a requirement for pilots flying air carrier aircraft, although a commercial pilot with instrument rating may fly as first officer (copilot) on an air carrier aircraft.

Aircraft type rating

Certain pilots are required to hold a type rating for flying large aircraft and small turbine-powered airplanes.

Chapter 1

Examples include the B-737 rating, DC-9 rating, and Learjet rating.

Type certificates are required before a manufacturer can produce and sell an aircraft and receive an airworthiness certificate.

Part 23—Airworthiness standards: Normal, utility, and acrobatic category airplanes

FAR part 23 contains detailed design and test requirements for the certification of small airplanes. As defined in FAR part 1, *small aircraft* means aircraft of 12,500 pounds (lb) or less, maximum certificated takeoff weight.

Part 25—Airworthiness standards: Transport category airplanes

FAR part 25 contains detailed design and test requirements for the certification of transport category airplanes.

Maintenance, preventive maintenance, rebuilding, and alteration (FAR part 43)

Maintenance means inspection, overhaul, repair, preservation, and the replacement of parts, but excludes preventive maintenance. *Preventive maintenance* means simple or minor preservation operations and the replacement of small standard parts not involving complex assembly operations.

Who can work on an aircraft?

Part 43.31 lists individuals and organizations that can perform maintenance: manufacturers, repair stations, air carriers, mechanics, and people working under the supervision of mechanics. It even allows pilots to work on aircraft. In reality, anyone can work on an aircraft, either by holding a certificate or working under supervision of someone holding a certificate.

Who can sign a logbook?
While anyone can work on an aircraft, parts 43.5 and 43.7 allow only a privileged few to "approve an aircraft for return to service." Just certified [aircraft and power plant (A&P)] mechanics, repair stations, manufacturers, air carriers, and pilots performing preventive maintenance can approve an aircraft for return to service. The real power in being an A&P mechanic is the authority given by the U.S. government to approve an aircraft for return to service.

The holder of a private or greater pilot certificate issued under FAR part 61 may perform preventive maintenance on any aircraft owned or operated by that pilot which is not used under part 121, 125, 127, or 135 (air taxi, air carrier, and all turbine-powered aircraft).

FAA Aircraft Operating Certificates

FAR part 121
This is an *air carrier certificate*. FAR part 121 operators are major airlines operating large aircraft with a passenger capacity of more than 30 seats or a payload of more than 7500 lb and engaged in interstate or overseas transportation for compensation or hire. The operator holds itself willing to furnish transportation (advertises) to the general public (called common carriage). FAR part 121 is a detailed document defining management requirements, record-keeping standards, operational requirements, and dispatching and flight release rules among other criteria. The FAA assigns airworthiness inspector(s) and principal

operating inspector(s) to each air carrier certificate holder for compliance and surveillance purposes.

FAR part 125
This part governs the operators of civil airplanes with a passenger capacity of 20 or more seats or a payload of 6000 lb or more when common carriage is not involved. Therefore, the operator does not hold itself out to the public (does not advertise). However, part 125 provides, under certain conditions, for the operator to fly for hire. Examples are large aircraft, like the Boeing 737, owned by a corporation for transportation of its own or customer personnel.

FAR part 127
This part applies to scheduled air carrier operations with helicopters.

FAR part 135
This part covers air taxi and some commuter operations.

FAR part 119
A relatively new regulation applies, in addition to operations already covered by FAR part 121, to all passenger-carrying operations for compensation or hire in airplanes with a capacity of 10 or more seats, excluding crew seats. This new rule effectively commits commuter operations to the same rules as FAR part 121 air carrier operations.

FAR Part 65—Certification of Airmen Other than Flight Crew
FAR part 65 includes certification of air traffic control tower operators, aircraft dispatchers, mechanics, and

parachute riggers. The following sections are relevant to the content of this book.

Aircraft dispatchers

Part 121 air carriers (major airlines) and certain FAR part 135 (commuter) operations require a certificated dispatcher to share responsibility with the pilot-in-command (PIC) in the operation and control of a flight. Dispatcher's duties shared with the PIC include weight and balance calculations, determining performance operating limitations, calculating takeoff criteria, and evaluating meteorological conditions.

Mechanics

Mechanics must pass comprehensive written, oral, and practical tests as well as have required experience and/or instruction in all phases of aircraft repair and maintenance in order to obtain an FAA mechanic certificate. The certificate can have either an airframe or power plant rating or both (A&P).

FAR part 91

FAR part 91, general operating rules, is the basic regulation on which other operating regulations are developed. For example, FAR part 121 is the air carrier regulation for operating large aircraft for compensation. FAR part 121 does not cover all operation subjects; therefore, FAR part 91 applies to fully comply with FAA requirements.

Some excerpts from FAR part 91 applicable to the subject of this book are as follows:

1. The pilot-in-command of a civil aircraft is responsible for determining whether that aircraft is in condition for safe flight.

2. An approved airplane flight manual containing operating limitations must be available for the flight crew.
3. Each pilot-in-command, before beginning a flight, shall become familiar with all available information concerning that flight.
4. Every civil aircraft must have a current airworthiness certificate and a registration certificate displayed at the cabin or cockpit entrance.

Inoperative instruments and equipment

Every turbine-powered aircraft must have on board an FAA-approved minimum equipment list (MEL) which authorizes an aircraft to continue flight with certain inoperable instruments or equipment under specific conditions and limitations. The use of, and the content of, the MEL is described in FAR part 91.213.

Note The MEL and its interpretation are probably the subject of greatest controversy among pilots, maintenance personnel, and aircraft operators. Without an MEL, many flights would have to be canceled, thus seriously limiting an airline's purpose of providing scheduled, reliable transportation.

Maintenance, preventive maintenance, and alterations

According to FAR part 91.903, the owner or operator of an aircraft is primarily responsible for maintaining that aircraft in an airworthy condition. The owner or operator of an aircraft must possess the maintenance manual(s) issued by the manufacturer of the aircraft.

Note The maintenance manuals for a large turbojet airplane occupy a 5-ft shelf.

Maintenance required, FAR part 91.405
This part generally refers to FAR part 43 which covers maintenance, preventive maintenance, rebuilding, and alteration.

Miscellaneous Definitions and Information

Crew complement—Air carrier aircraft
Most late-model airline turbojet aircraft require a crew of two: captain (pilot-in-command) and first officer (F/O). Some older-design aircraft still in service, such as the Boeing 727, require a crew of three: captain, first officer, and second officer (flight engineer).

In addition, a cabin crew (flight attendants, F/A), who are trained in emergency procedures, is required. The number of flight attendants required is dependent on the number of passenger seats.

Nongovernmental organizations
The International Civil Aviation Organization (ICAO) is a specialized agency of the United Nations whose objective is to develop the principles and techniques of international air navigation and to foster planning and development of international civil air transport.

The Aircraft Owners and Pilots Association (AOPA) is a membership organization of general-aviation owners and pilots. The Air Transport Association (ATA) is an organization of representatives from major airlines. The Airline Pilot's Association (ALPA) is an airline pilots' union. Members are airline pilots from most major airlines except American Airlines. American Airlines pilots have their own union: the Allied Pilots Association (APA). The National Business Aircraft Association

(NBAA) members are representatives of operators of business aircraft.

Note AOPA, ATA, ALPA, and NBAA are special interest groups. Some of these groups have lobbies and can exert pressure on Congress as well as the FAA.

Aircraft operating information
All aircraft must have onboard the following:
1. Aircraft manual containing normal, abnormal, and emergency procedures
2. Cockpit checklist
3. Aircraft logbook
4. Minimum equipment list

FAR part 91.503 includes detailed requirements.

Ferry flight
A ferry flight is a nonrevenue flight for the following purposes:
1. To return an aircraft to base
2. To deliver an aircraft from one location to another
3. To move an aircraft to and from a maintenance base. Ferry flights, under certain conditions, may be conducted under terms of a special flight permit.

Scheduled maintenance
Air carrier and all turbine-powered aircraft undergo A, B, C, and D checks at scheduled flight times. During an A check, operation of all lights and brakes is checked, fluids are replenished, and tires are checked. The B, C, and D checks are progressively more thorough, with a D check being a complete overhaul.

General-aviation aircraft require a 100-hour inspection and an annual inspection.

Turbojet engine operation—Power settings

The power (thrust) of a turbojet engine is determined by the engine pressure ratio (EPR). This is the ratio of the compressor inlet pressure to the turbine outlet pressure.

Compressor-turbine rotor speed is presented in percent rpm, N_1 and/or N_2. A single-rotor engine rpm is N_1; a dual-rotor engine rpm is indicated by N_1 and N_2.

Exhaust gas temperature (EGT) is an indication of the turbine outlet temperature. EGT is a critical temperature. Excessive EGT can damage the turbine.

Crew resource management (CRM)

CRM has become a buzzword, especially in airline pilot operations. CRM involves the management and utilization of all the people, equipment, and information in the aircraft. In other words, CRM applies to any activity involving a team, equipment, and a dynamic situation. Most airlines now include CRM in their training programs.

The following chapters consist of ASRS reports submitted to NASA by pilots who were confronted with maintenance-related problems.

The last page of each chapter includes a summary, assessment, and lessons learned from an evaluation of the chapter's ASRS reports.

2

Problems with MEL (Minimum Equipment List)

Every turbine-powered aircraft must have available (usually onboard) an FAA-approved minimum equipment list which authorizes an aircraft to continue flight with certain inoperable instruments or equipment under specific conditions and limitations. The use of, and content of, the MEL is described in FAR part 91.213.

The highlights of FAR part 92.213 are as follows:

The inoperative instruments and equipment must be removed from the aircraft, the cockpit control placarded, and the maintenance recorded in accordance with FAR part 43.9 (maintenance FAR).

If the inoperative instruments are not removed from the aircraft, they must be deactivated and placarded "inoperative." If deactivation involves maintenance, it must be recorded in accordance with FAR part 43.

Furthermore, a determination must be made by a pilot who is certificated and appropriately rated under FAR part 61, or by a person who is certificated and appropriately rated to perform maintenance on the aircraft, that the inoperative instrument does not constitute a hazard to the aircraft due to removal or deactivation.

The aircraft records available to the pilot must include an entry describing the inoperable instruments and equipment. The aircraft must be operated under all applicable conditions and limitations contained in the MEL. Certain instruments and equipment may not be included in an MEL as defined in FAR part 91.213.

As discussed in Chap. 1, the MEL and its interpretation are probably the subject of greatest controversy among pilots, maintenance personnel, and aircraft operators. However, without an MEL, many flights would have to be canceled, thus seriously limiting an airline's purpose of providing scheduled, reliable transportation.

After the 16 ASRS reports, a brief summary and analysis will follow, but a more complete overall evaluation is presented in Chap. 10.

Procedures Omitted for Disconnecting Constant-Speed Drive (ASRS Report #427408)

Date:	1999/02
Type of operation:	Air carrier, passenger flight
Aircraft:	Low-wing, medium-large transport, between 60,001 and 150,000 lb
	Engines: Two turbojets
	Number of crew: Two
Flight crew:	ASRS reporter: captain, pilot-in-command (PIC). 10,000

Problems with MEL (Minimum Equipment List)

hours, 240 hours within the past 90 days, 1400 hours in make/model. Air transport certificate, certificated flight instructor.

Narrative (unedited, ASRS reporter's own words)

Our aircraft had a maintenance carry-over item (deferred per MEL) (minimum equipment list). #1 generator was placarded inoperative. We researched the requirements in our MEL. In part, the MEL required the constant speed drive to be disconnected and verified disconnected by maintenance. The logbook stated that MEL items were complied with by maintenance. After takeoff, we noticed high oil temperature light (associated with the #1 constant speed drive) illuminate.

We called company and were advised to try the constant speed drive disconnect switch. When we did, we immediately got the constant speed drive low pressure light, followed by a drop in constant speed drive oil temperature. The constant speed drive apparently had not been fully disconnected, putting us at risk of fire. The flight continued without further incident. Contract maintenance worked on the aircraft after it was secured, and flight crew departed for hotel preventing verbal description of problem. Also, contract maintenance is not as familiar with company procedures as company maintenance.

Report synopsis

A B737 was dispatched in noncompliance with the #1 generator deferred as inoperative per the MEL, but special procedures for disconnecting the constant-speed drive were omitted.

Aircraft involvement
There was an aircraft equipment problem, nonadherence to legal requirements and published procedures, and nonadherence to a FAR requirement.

Cockpit Voice Recorder Inoperative and Deferred (ASRS Report #90653)

Date: 1998/07
Type of operation: Air carrier, passenger flight
Aircraft: Low-wing, large transport, between 150,001 and 300,000 lb.
Engines: Three turbojets
Number of crew: Three
Flight crew: ASRS reporter: second officer, flight engineer. 2800 hours, 180 hours within past 90 days, 1500 hours in make/model. Air transport certificate, flight engineer rating.

Narrative (unedited, ASRS reporter's own words)
Company issued deferred maintenance number for inoperative cockpit voice recorder and requested myself and other crew members to take the aircraft. Given the revised MEL/CDL (minimum equipment list/configuration deviation list) and the fact company MEL/CDL lists have not been updated, we were unable to

determine the legality of this situation. Since we were departing MIA (Miami, FL), a maintenance base, we have some doubts.

Report synopsis
Air carrier large transport flight crew question the legality of dispatch from major maintenance base with cockpit voice recorder inoperative and deferred.

Aircraft involvement
Aircraft was involved in an anomaly (an unsafe or illegal event).

Inoperative Fuel Gauge per MEL (ASRS Report #80626)

Date:	1988/01
Type of operation:	Air carrier, passenger flight
Aircraft:	Low-wing, large transport, between 150,001 and 300,000 lb
	Engines: Three turbojets
	Number of crew: Three
Flight crew:	ASRS reporter: captain, pilot-in-command. 10,000 hours, 150 hours within past 90 days, 150 hours in make/model. Air transport certificate, flight engineer rating, certified flight instructor.

Narrative (unedited, ASRS reporter's own words)
We left St. Louis en route to La Guardia with an inop-

erative #2 fuel quantity gauge per MEL (minimum equipment list). Thus dripsticks [sic] were used by maintenance at St. Louis to verify fuel in tank #2. In cruise flight, boost pump low pressure lights coming on told us we had very little fuel in tank #2. We proceed to La Guardia, landing uneventfully while crossfeeding #2 engine from tanks #1 and #3. Tank #2 was virtually empty on landing. Maintenance at St. Louis apparently did not put fuel in the tank, even though they filled out paperwork to the contrary. How to fix the problem? Fix the airplanes when something breaks and hire people who know what they are doing.

Report synopsis
Aircraft left with inoperative fuel gauge on one tank. Maintenance did not fill tank as reported.

Aircraft involvement
Aircraft was involved in an anomaly (an unsafe or illegal event).

Departure without Legally Required Documents (ASRS Report #300991)

Date:	1995/03
Type of operation:	Commuter air carrier, passenger flight
Aircraft:	Low-wing, light transport, between 14,501 and 30,000 lb
	Engines: Two turboprops

Problems with MEL (Minimum Equipment List)

Flight crew:
Number of crew: Two
ASRS reporter: captain, pilot-in-command. 15,200 hours, 200 hours within past 90 days, 8500 hours in make/model. Air transport certificate.

Narrative (unedited, ASRS reporter's own words)

On a turnaround, with several mechanical problems, I was performing the required items to operate, per MEL (minimum equipment list). A company maintenance representative was deferring several items. I was mainly concerned with sticking the tank (we had an inoperative fuel gauge), and calculating fuel. The maintenance representative had removed the MEL document from the aircraft without my knowledge after we departed (pressed for time). We were told that they would send the MEL up on the next flight and that they would have another one at the hub. The MEL contained the weight and balance documents as well. The contributing factor, here, is a bonus program for on-time departures and the station involved, which classifies all late departures as crew infractions. At this airline, the tail wags the dog.

The chief pilot inquired as to how I had complied with the MEL without it on board [*sic*]. I had reviewed it inbound, and was working on all compliance when it was removed. As captain there are only so many things I can do at one time. The F/O (First Officer) was essentially useless. He is a "paid for training" foreign national here to build up flight time and only good for consuming peanuts and soda pop. I do not know who removed the document—the maintenance representative or the F/O. The chief pilot has instructed the Director of Maintenance to tell all mechanics never remove the MEL document from the aircraft. Only the aircraft log needs signatures from maintenance entries. The aircraft log was reviewed by us, prior to departure. We just did not check the other documents which should not have been disturbed. On arrival at the hub, we reconstructed the manuals and weight and balance data (rebuilt from files), and with all legal documents on board with no delay, on next departure.

Supplemental information We landed at Lafayette, LA with the 2 open "write-ups." The mechanic at the base deferred the 2 items. He removed the maintenance book and the MEL from the aircraft. When he had called in and taken care of the deferrals on the phone, he failed to put the MEL back on the aircraft, with the appropriate maintenance papers.

Report synopsis
A commuter aircraft, which had maintenance-deferred items (per aircraft's MEL), departed without legally required documents onboard.

Aircraft involvement
Aircraft was involved in an anomaly (an unsafe or illegal event).

Autopilot Altitude Hold Feature Inoperative (ASRS Report #104365)

Date:	1989/02
Type of operation:	Air carrier, passenger flight
Aircraft:	Low-wing, medium-large

Problems with MEL (Minimum Equipment List)

	transport, between 60,001 and 150,000 lb
	Engines: Two turbojets
	Number of crew: Two
Flight Crew:	ASRS reporter: first officer (F/O). 3400 hours, 210 hours within the past 90 days, 1600 hours in make/model. Air transport certificate, flight engineer rating.

Narrative (unedited, ASRS reporter's own words)

This event involved an ambiguity in the company's MEL (minimum equipment list) for the medium large transport aircraft. On this particular trip, the aircraft had a previous write up in the E6 log book which stated that the altitude hold caused an oscillation in the control column with either the #1 or #2 autopilot selected. Repair of this discrepancy was deferred and a placard was placed in the E6 log book stating "altitude hold #1 and #2 inoperative as per MEL." The altitude hold button on this autopilot control panel was also placarded as inoperative. The LMP (landing mode performance) status remained category III. The company MEL states simply not to use the altitude hold mode in this situation. (No mention was made not to use the autopilot). A check of the MEL under "perf" section specifically states that any part of the system not listed as inoperative may be used.

The initial write-ups, the sign-off and the MEL system description did not specifically prohibit using the autopilot. We therefore decided the use of the autopilot in the perf mode was acceptable and legal. An FAA maintenance inspector who was riding the jump seat later questioned the use of the autopilot in light of this discrepancy. If the FAA inspector is correct in this matter, then it seems that the company MEL should be rewritten to specifically prohibit the use of all autopilot modes in this situation. Additionally, if use of the autopilot is inappropriate, then the company MEL and maintenance manual should dictate that the LMP status of the aircraft be downgraded to category I only.

Report synopsis
The FAA maintenance inspector wondered if it was OK to use autopilot with the altitude hold feature placarded inoperative.

Aircraft involvement
Aircraft was involved in an anomaly (an unsafe or illegal event).

MEL Item Not Repaired or Written Off in Log Book (ASRS Report #102264)

Date:	1989/01
Type of operation:	Air carrier, passenger flight
Aircraft:	Low-wing, medium-large transport, between 60,001 and 150,000 lb
	Engines: Two turbojets
	Number of crew: Two
Flight crew:	ASRS reporter: captain, pilot-in-command. 8700 hours, 250 hours within the past 90

Problems with MEL (Minimum Equipment List)

days, 6500 hours in make/model.

Narrative (unedited, ASRS reporter's own words)

I reported to the aircraft at MCI (Kansas City, MO) after an overnight to operate from MCI-OKC-HOU (Kansas City–Oklahoma City–Houston). Aircraft had a MEL (minimum equipment list) item open in the logbook. A single channel of the pitch trim had been written up as inoperative and OK to continue per MEL, and a green sticker was on the pitch trim indicator light flight operations manual.

After airborne en route to OKC I thought to myself that it was odd that a green sticker was still on the light, because the procedure is to change the green sticker to a white sticker (to be done by maintenance) when passing through a maintenance base. The FOM (flight operating manual) specifically stated that if you pass through a maintenance base with a green sticker, you will either get maintenance to clear the problem or change it to a white sticker and enter the item in the DMI page. After going back to the logbook, I realized the airplane had been through twice the day before, and the item had not been transferred by maintenance to the DME page and a white sticker applied. Here I was flying an airplane in violation of the procedures in the flight operations manual. I decided to fly the airplane on to OKC. At OKC maintenance was summoned and told the story. The mechanic decided to go back 2 log pages and signed the discrepancy off as "system operates OK as reported by pilots, OK for flight" and dated it the day before (it was working OK). So I was off the hook and also the pilot before me who forgot to get the item

30 Chapter 2

transferred to the DMI page. If the mechanic had not done what he did but cleared it, I'm sure I would have been violated.

Report synopsis
A minimum equipment list item was not repaired or written off in the logbook when the aircraft went through a repair station. Reporter didn't notice it until en route. At next maintenance base stop, the logbook was changed (to be legal).

Aircraft involvement
Aircraft was involved in an anomaly (an unsafe or illegal event).

Delay Due to Conflicting Interpretation of MEL by Maintenance (ASRS Report #82374)

Date:	1988/02
Type of operation:	Air carrier, passenger flight
Aircraft:	Low-wing, wide-body transport, over 300,000 lb
	Engines: Three turbojets
	Number of crew: Three
Flight crew:	ASRS reporter: captain, pilot-in-command. 15,000 hours, 200 hours within the past 90 days, 200 hours in make/model. Air transport certificate, flight engineer rating.

Problems with MEL (Minimum Equipment List)

Narrative (unedited, ASRS reporter's own words)

In attempting to operate a wide body aircraft flight out of ORD (Chicago) in February, a series of very complicated and involved operational issues surfaced which caused extensive delays and near cancellation of our flight. These issues principally involved MEL (minimum equipment list) items and inoperative components with regard to operations into icing conditions. Of major concern was the difficulty of interpreting the MEL to determine if the operation was permissible. This was further complicated by conflicting MEL directives as contained in computer storage that engineering has access to versus information that maintenance provided via its maintenance release document. The step by step resolution of this problem by the flight crew, maintenance and dispatcher was undertaken, and what appeared to be a problem that would be resolved momentarily continued and became more complicated and time consuming. These were the factors, information and problems provided to an interested audience. The following highlights additional details or concerns and problems regarding this flight.

Operations were delayed due to snow and icing conditions at ORD. Light snow was falling and the temperature was 26 degrees F. The aircraft was initially deiced some 10–15 minutes prior to scheduled departure time. Then passenger service announced a departure delay from the scheduled XA05 to a new time of XA25. As snow continued to fall, coupled with this long delay, another deice service was requested. This was accomplished and conducted at XA50 and our flight departed the blocks at XA55. Departure delays were occurring and taxi was slow. Approximately 25 minutes after block departure we decided to return to the gate for additional deicing. We were nearing #1 position for takeoff, but snow was still falling and it could not be assured that the wings/control surfaces were free of frost, snow or ice.

About this time our dispatcher contacted us regarding some question as to the possibility that flight into known icing conditions may not be appropriate. We had a #1 pneumatic temperature gauge that was inoperative, but operation under MEL authority. This seemed appropriate so long as the systems supplying pneumatic air for ice protection had operational temperature gauges. However, his point was well taken as further discussion proved that the MEL in this area was far from being clear. Was the system with the inoperative temperature gauge to be turned off if the flight was to be operated into icing conditions? Then what considerations should be given to reduced pneumatic capacity? Should the #1 pneumatic system have been deactivated by maintenance under this situation?

At this time engineering was contacted and when he pulled up the MEL from his computer in San Francisco, it indicated that operation into icing conditions was permitted. We were in agreement, and I asked that a copy of maintenance release document with the MEL information be sent to me through computer terminal at the gate and we would be underway. I received the document and proceeded to board the aircraft when to my shock I noticed the same restrictions were still part of the document: Flight shall not be made into icing conditions. So, back to the gate telephone and another call. In the meantime, we patched in the dispatcher and had ORD line maintenance in the communication loop. We were all perplexed! I was satisfied that the operation would have been correct and proper as ORD maintenance and CHI (Chicago dispatch), but I explained that it would be necessary to have correct paperwork to

Problems with MEL (Minimum Equipment List) 33

show this as part of the aircraft logbook documentation.
  This problem took nearly one hour to resolve! It was finally decided that maintenance should deactivate the pneumatic system with the inoperative temperature gauge. Thus, a new MEL would come into play authorizing operations into icing conditions. This complicated and round-about way to operate under MEL authorization should not exist. Ironically, after takeoff, a wing anti-ice valve failed to open. Since we were in icing conditions, we continued to climb to a high altitude where we were clear of ice accumulation. Our S/O (Second Officer) could not open the valve after several attempts by using the irregular procedures, and with the assistance of our dispatcher, we were prepared to fly to a new destination that was free of icing conditions. However, one last attempt at opening the valve succeeded, and we flew on to our original destination where the aircraft was taken out of service.

Report synopsis
Wide-body aircraft was delayed because of inoperative gauge due to conflicting interpretation by maintenance of the minimum equipment list.

Aircraft involvement
Aircraft was involved in an anomaly (an unsafe or illegal event).

Non-MEL Item Was Written Up as MEL Item (ASRS Report #83898)

Date: 1988/03
Type of operation: Air carrier, passenger flight
Aircraft: Low-wing, medium-large

	transport, between 60,001 and 150,000 lb
	Engines: Two turbojets
	Number of crew: Two
Flight crew:	ASRS reporter: captain, pilot-in-command. 11,000 hours, 180 hours within the past 90 days, 9000 hours in make/model. Air transport certificate, certified flight instructor.

Narrative (unedited, ASRS reporter's own words)

I will describe an event where we complied with a MEL (minimum equipment list) procedure to the letter of the law and still flew a dangerous aircraft. I reported to an aircraft for flight from DTA–ATL (Detroit Metropolitan, MI–Atlanta, GA) after another aircraft had been maintenance cancelled, a medium large aircraft. After checking the logbook, a write up existed that the elevator power annunciator light would not come on when full elevator down was put in on the yoke. Maintenance checked it out and informed me they would MEL it. So, in accordance with the MEL, we turned off all hydraulic pumps and depleted pressure with the spoilers. When the F/O (first officer) pushed the yoke full forward, the mechanic and I both visually checked the movement of the elevator down—it was not much movement, but the mechanic informed me that was normal. We returned hydraulic pressure and checked again. Same movement—all was normal.

We departed into snow and icing conditions to ATL. The MEL limit with a full airplane. Upon arrival at ATL, maintenance met the aircraft, with the switch believed

Problems with MEL (Minimum Equipment List) 35

to be the problem, after DTW (Detroit, MI) maintenance called them. Before changing the switch for the light, a mechanic climbed up in the tail and checked movement and pressure at the elevator. It was not normal. They bled the accumulators and got the light to come on dimly. The problem was low pressure not giving full deflection of the elevator. I had just flown a dangerous aircraft with no hydraulic boost in the elevator.

We changed aircraft and returned to DTW. I believe the MEL should be changed completely. If the elevator power light will not test, the aircraft should be grounded till [sic] repaired, not a visual inspection of elevator movement. If we did not go to ATL, this aircraft could have flown for months with procedures for MEL. You have no way of telling if you have full elevator travel from a visual inspection on the ground. I don't know if this is the appropriate place to report this, but I have had great success with your system in the past for flight problems. With the economic pressures of major airlines today, we need this system to keep air travel safe. It's dangerously close to unsafe today thanks to today's pressure for the bottom line.

Note Callback conversation with the reporter revealed the following: Reporter turned in report to his company which also is copied to the ALPA (Airline Pilot's Association) safety committee, so they are looking at the problem. He also stated [that] mechanics he has talked to agree [that] looking at horizontal stabilizer is not satisfactory to ascertain if system [is] working properly.

Report synopsis

Air carrier, medium-large transport was dispatched under MEL. Problem was later found to be non-MEL item, and MEL list maintenance verification procedure is apparently inadequate.

Aircraft involvement
Aircraft was involved in an anomaly (an unsafe or illegal event).

Manufacturer Confirms Emergency Lights Not Deferrable (ASRS Report #106380)

Date:	1989/03
Type of operation:	Commuter air carrier, passenger flight
Aircraft:	Low-wing light transport, between 14,501 and 30,000 lb
	Engines: Two turboprops
	Number of crew: Three
Flight crew:	ASRS reporter: captain, pilot-in-command. 4400 hours, 200 hours within the past 90 days, 300 hours in make/model. Air transport certificate.

Narrative (unedited, ASRS reporter's own words)

The airplane had a deferred item "Emerg. Lights." I took the MEL (minimum equipment list) list to dispatch and maintenance and showed them that the MEL did not state anything about emergency lights. They said it was part of the cockpit lighting which was deferrable. Later in the week, maintenance found out from manufacturer that the emergency lights were not deferrable. Seven days later I was told by chief pilot's office the lights were not deferrable.

Problems with MEL (Minimum Equipment List) 37

Report synopsis
The PIC of commuter light transport questioned the MEL requirement. Maintenance and dispatch confirmed that emergency lights could be deferred. Manufacturer later denied this and confirmed that emergency lights were not deferrable.

Aircraft involvement
Aircraft was involved in an anomaly (an unsafe or illegal event).

Gear Door Removed but Weight and Speed Restrictions Not Listed (ASRS Report #426491)

Date:	1999/01
Type of operation:	Commuter air carrier, passenger flight
Aircraft:	Low-wing, medium transport, between 30,001 and 60,000 lb
	Engines: Two turbojets
	Number of crew: Two
Flight crew:	ASRS reporter: captain, pilot-in-command. 11,340 hours, 234 hours within the past 90 days, 3820 hours in make/model. Air transport certificate.

Narrative (unedited, ASRS reporter's own words)
Reporting one dark morning at the hangar, I found my F/O (First Officer) on the aircraft saying she had pre-flighted and was ready to go with acceptance checklist

complete. I checked through the aircraft clipboard and noticed in the deferral that "right main landing gear door bellcrank has play, OK per configuration deviation list." After repositioning to the gate, a look at our dispatch release showed us flight planned using our normal climb, cruise and descent profiles at normal airspeeds, as high as 320 knots indicated air speed (KIAS), and the ETE (estimated time en route) looked normal. In other words, everything seemed normal to me at this point.

About two-thirds into our flight, my F/O mentioned some weight penalties noted on the release. There were two sets of penalties and restrictions, both for the same configuration deviation list item in the deferred log. One set for missing gear door brushes and another for the gear doors themselves. I asked her if it was the brushes that were missing, because surely she would've noticed if the gear door was missing, right? And surely dispatch would have planned the flight for 250 KIAS, along with a greater ETE and fuel burn that would've stood out as unusual, right? At least that's what we thought. In order to make sure, we called maintenance on the radio, and to our shock, the right gear door had in fact been removed, and she had missed it in her preflight. We slowed to 250 KIAS for the remainder of the flight, and made a write-up in the clipboard so that an inspection of the gear could be made. No damage was noted upon inspection.

While still en route, we called our dispatcher, and they admitted that this was a recurring problem that sometimes slips through. In this case, as in many incidents, it's a string of errors from different sources that leads to this outcome. I realize the PIC (pilot-in-command) has the final responsibility for the proper operation of the aircraft, but I need help in achieving that. While it seems unlikely that something like this could be missed, it's a recurring problem. Of course we must avoid complacency during preflight.

Problems with MEL (Minimum Equipment List) 39

Also, a few other things that would help: 1) A large placard placed on each side of the cockpit instrument panel, under the existing speed limitations placard, stating "gear doors removed, maximum 250 KIAS." 2) Make sure the gear doors and door brushes [are] separate configuration deviation list items, not bumped together under one entry as they are now with the 250 KIAS limitation a small note at the bottom of it all. 3) Required a clear unambiguous entry in the deferral log: "Gear Doors Removed, Max 250 KIAS!" 4) A better means of flagging aircraft with this deferral in the dispatcher's computer, even programming it to prevent flight planning with normal speed profiles. 5) Shortening the expiration of this configuration deviation list item from 7 days to 3 days would limit the risk of exposure to this sort of problem.

Report synopsis
A Canadian CL65 was dispatched in noncompliance with a major gear door removed per the configuration deviation list. Required weight and speed restrictions were not listed in the dispatch release nor was the cockpit instrument panel placarded.

Note The configuration deviation list is the same as the MEL.

Aircraft involvement
Aircraft was involved in an anomaly (an unsafe or illegal event).

Landing Gear Door Positioned Improperly (ASRS Report #87619)

Date: 1988/05

40 Chapter 2

Type of operation:	Air carrier, passenger flight
Aircraft:	High-wing, medium-large transport, between 60,001 and 150,000 lb
	Engines: Four turbojets
	Number of crew: Two
Flight crew:	ASRS reporter: first officer (F/O). 6000 hours, 100 hours within the past 90 days. Commercial license, instrument rating.

Narrative: (unedited, ASRS reporter's own words)

Upon receiving aircraft for first flight of the day, a normal preflight inspection was conducted by the F/O (first officer). F/O noticed maintenance was working on the R/H main landing gear brake fan. Passengers were boarded and maintenance advised myself that they were going to MEL (minimum equipment list) the brake fan. The logbook was signed and aircraft released from maintenance. Everything was normal until the gear was selected up after takeoff. The R/H gear (red light) not in position selected, light remained on. The abnormal checklist was consulted and crew elected to return to the field.

Aircraft returned to the gate where maintenance inspected the R/H gear and found that they forgot to secure the gear door in the correct position. The mechanics involved were experienced people who were rushed to get the flight out on time. As the gear was retracted, but the gear door was not, aircraft performance was not significantly effected [sic]. However, it is the emphasis to get the flight out on time that may be detracting from the safety and thoroughness of maintenance procedures. To prevent this particular occurrence

Problems with MEL (Minimum Equipment List) 41

from happening again, I am determined to re-check maintenance's positioning of the gear doors whenever work has occurred on the tires or brakes.

Report synopsis
Landing gear door was positioned improperly by airline maintenance.

Aircraft involved
Aircraft was involved in an anomaly (an unsafe or illegal event).

Wingtip Light Inoperative and Deferred in Noncompliance with MEL (ASRS Report #425630)

Date:	1991/01
Type of operation:	Air carrier, freight
Aircraft:	Low-wing, large transport, between 150,001 and 300,000 lb
	Engines: Three turbojets
	Number of crew: Three
Flight crew:	ASRS reporter: captain, pilot-in-command. 7600 hours, 130 hours within the past 90 days, 2500 hours in make/model. Air transport certificate, flight engineer rating, certified flight instructor.

Narrative (unedited, in ASRS reporter's own words)
We were operating a scheduled Part 121 cargo flight

from Vienna, Austria to Cologne, Germany with an en route stop at Nuremberg, Germany. During his preflight inspection on the ramp at VIE, my S/O (second officer) informed me that the white position light located on the L-hand wingtip (facing rearward) was inoperative. I was very busy with several things involving deice coordination, en route clearances, on-time departure concerns, etc. I considered this particular F/E to be extremely competent and so I requested him to review the MEL (minimum equipment list) and brief me on go/no-go status. He did so, and informed me that this item could be MEL'd. A contract mechanic was in the cockpit, and he agreed with this and proceeded to enter a sign-off in the logbook for this item, referencing the MEL.

I did not personally read the MEL reference as has been my habit, but accepted the position of the F/E and mechanic. Our company procedure allows us to depart without a control number from maintenance control and get the number en route, if the problem is discovered within 30 minutes of departure, and all other logbook/MEL procedures have been complied with. We departed for Nuremberg and requested the item control number via company radio [and were told that] this item was a no-go item for the MEL for night ops. We were told to overfly Nuremberg and proceeded to Cologne, as no parts were available in Nuremberg. Result of this was that I operated the aircraft in an unairworthy condition and caused the company to have to dispatch an additional aircraft to Nuremberg.

Report synopsis

A B727 was dispatched with the left wing aft light inoperative and deferred in noncompliance with MEL.

Aircraft involvement

Problems with MEL (Minimum Equipment List) *43*

Aircraft was involved in an anomaly (an unsafe or illegal event).

Window Cracked—Not Addressed by MEL (ASRS Report #122950)

Date:	1989/09
Type of operation:	Air carrier, passenger flight
Aircraft:	Low-wing, medium-large transport, between 60,001 and 150,000 lb
	Engines: Two turbojets
	Number of crew: Two
Flight crew:	ASRS reporter: captain, pilot-in-command. 9000 hours, 270 hours within the past 90 days, 7000 hours in make/model. Air transport certificate.

Narrative (unedited, ASRS reporter's own words)

At RSW (Southwest Florida Intl.), outer pane of F/O (first officer) window cracked while aircraft was parked at gate. Verified that only outer pane was cracked and that structural integrity of window was not effected [*sic*]. Crack did not obstruct F/O line of vision. Maintenance and dispatch said I could go with MEL (minimum equipment list) for window heat inoperative. I flew at 25000 feet and airspeed at 250 knots. In retrospect, I feel I was not legal to go since MEL did not specifically address a cracked windshield.

Report synopsis

An air carrier medium-large transport departed on scheduled flight with the first officer's windshield cracked. Maintenance had verified that equipment list restrictions were complied with.

Aircraft involvement
Aircraft was involved in an anomaly (an unsafe or illegal event).

MEL Operating Weight Restrictions Not Applied Due to Software Failure (ASRS Report #427920)

Date:	1999/02
Type of operation:	Air carrier, passenger flight
Aircraft:	Medium-large, low-wing transport, between 60,001 and 150,000 lb
	Engines: Two turbojets
	Number of crew: Two
Flight crew:	Captain, pilot-in-command.
ASRS reporter:	Dispatcher

Narrative (unedited, ASRS reporter's own words)
Received my briefing from late night dispatcher. No mention of anti-skid MEL (minimum equipment list) for my aircraft. I calculated the flight plan. The MEL never appeared on the release. I released the flight and had no communication from the crew. I received a call at home from the dispatcher who relieved me. He had received a call from the crew on the next leg of flight. They ques-

Problems with MEL (Minimum Equipment List) 45

tioned why the MEL was not on their flight plan. After inquiring with maintenance, it was determined that an MEL did exist and should have been on the flight plan.

Performance penalties should have been applied and were not because I was unaware of MEL. The crew should have questioned the release but did not. The next crew lead [*sic*] to our discovery of the computer problem. Our maintenance computer did not send correct information to our new flight computer system. (My flight was only one that had a problem). Management was advised, and now we have a "backup" system and we must verify each MEL with Maintenance.

Report synopsis
MEL restrictions were not applied to aircraft operating weights due to computer software failure.

Aircraft involvement
Aircraft was involved in an anomaly (an unsafe or illegal event).

Flight Dispatched in Noncompliance with Open, Unanswered Log Report on Altimeter (ASRS Report #426500)

Date:	1999/01
Type of operation:	Air carrier, passenger flight
Aircraft:	Low-wing, medium-large transport, between 60,001 and 150,000 lb
	Engines: Two turbojets
	Number of crew: Two

Flight crew: ASRS reporter: first officer. Commercial license with instrument rating.

Narrative (unedited, ASRS reporter's own words)

We were one hour out from LAX (Los Angeles) on our DFW-LAX (Dallas-Fort Worth—Los Angeles) flight and noticed a mechanical problem with the captain's altimeter. As this was our aircraft for the return flight to DFW, some wanted it fixed at LAX. So we sent the ACARS message to maintenance and started the logbook entry.

Before the logbook entry was completed, a passenger medical emergency came to my attention that required diversion to PHX (Phoenix). The logbook entry was not completed nor maintenance performed until we reached LAX. Post flight review of the MEL (minimum equipment list) told me that the mechanical discrepancy should have been recorded and either placarded or repaired at the point of first landing at PHX instead of flight completion at LAX.

Report synopsis

An MD Super 80 was dispatched in noncompliance with an open unanswered log report on the captain's altimeter.

Aircraft involvement

Aircraft was involved in an anomaly (an unsafe or illegal event).

Inoperative Fuel Gauge and Incorrect Dripstick Fuel Quantity

Problems with MEL (Minimum Equipment List)

(ASRS Report #429125)

Date: 1999/02
Type of operation: Commuter air carrier, passenger flight
Aircraft: High-wing, medium transport between 30,001 and 60,000 lb
Engines: Two turboprops
Number of crew: Two
Flight crew: ASRS reporter: first officer. 2200 hours, 50 hours within the past 90 days, 1000 hours in make/model. Commercial license.

Narrative (unedited, ASRS reporter's own words)

This report involves an MEL (minimum equipment list) on the ATR (aircraft type) regarding inoperative fuel gauges. The MEL allows the airplane to be flown for up to ten days with an inoperative fuel quantity indicator. The alternate system used to determine the actual fuel on board is highly inaccurate. We were given an airplane to take from EWR (Newark) to (BWI) Baltimore and the right fuel tank gauge was MEL'd. With help from maintenance we checked the dripstick [*sic*] numbers, checked the chart and then determined we had 2100 lbs. in the right tank.

We landed at BWI having burned 600 lbs. from the right tank. At this point we should have had 1500 lbs. left. On the ground at BWI we added 80 gallons (536 lbs.). At this point the captain and I checked our dripsticks, and we both checked the numbers and the charts twice and the sticks indicated just over 1000 lbs. This was a 1000 lb. discrepancy in what the fuel sticks said we had and what we thought we had. So we added another 100 gallons (670 lbs.) and "resticked" the tanks, and now the drip tanks indicated 2100 lbs. Another 400 lb. discrepancy. This system is dangerous, because you cannot accurately determine the amount of fuel on board. If you have a situation where you cannot take a large amount of fuel because of a high payload, and you have 1000 lbs. less than you thought you had, you stand a good chance of running out of fuel. The MEL, if allowed to exist at all, should be limited to one flight to a station where repairs can be made.

Note Callback conversation with reporter revealed the following information: The reporter stated the problem with the fueling was mainly caused by attempting to fuel using dripsticks to verify the load and having the wrong stick or sticks installed in the tank. The reporter said the metered fuel from the truck was being used to fuel both tanks and should have been used for only the right tank until the correct load was metered into the right tank.

Report synopsis
An ATR42 was dispatched with the right fuel tank quantity indicator deferred as inoperative, per the MEL, and the wrong dripstick or dripsticks located in the tank.

Aircraft involvement
Aircraft was involved in an anomaly (an unsafe or illegal event).

Summary and Assessment: Problems with MEL
This chapter includes 16 ASRS reports from 1988 through

Problems with MEL (Minimum Equipment List) 49

1999 regarding MEL problems. Although each of the 16 reports involved different aircraft systems or components, each resulted in an illegal and/or unsafe flight.

An MEL is prepared by the aircraft manufacturer for an FAA-approved aircraft design, and copies are made available to all maintenance departments as well as being part of each individual aircraft's paperwork and carried in the aircraft.

In accordance with the MEL, an aircraft can be dispatched with certain components and/or systems inoperative, normally with flight restrictions. These deferred items must be repaired within a certain period.

As gleaned from the preceding ASRS pilot reports, the following MEL problems are revealed:

1. Through misinterpretation of the MEL, a problem was considered an MEL item when it was not (ASRS reports #83898, #106380, #90653, and #122950).

2. MEL procedures were incomplete (ASRS reports #427408, #80626, and #429125).

3. MEL items were not repaired or written off at repair/maintenance stations. PIC didn't notice until en route (ASRS reports #426500 and #102264).

4. Flight crew did not know of, or adhere to, MEL restrictions (ASRS reports #425630, #426491, and #427920).

5. Flight crew failed to visually check or missed maintenance MEL'd item on preflight (ASRS report #426491).

6. MEL document was illegally removed from aircraft. Flight crew did not notice until en route (ASRS report #300991).

Contributing factors pertaining to MEL problems are as

follows:

1. There is excessive pressure on maintenance personnel and flight crew for on-time departure.
2. PIC is distracted with other preflight tasks.
3. There is no company maintenance at the airport. This is more of a problem with commuter airlines.
4. PIC accepts maintenance's assessment of MEL procedures without personally checking.

Only maintenance can perform MEL procedures. A pilot cannot perform maintenance or repair, including preventive maintenance, on a turbine-powered aircraft. However, in accordance with FAR part 91.7, the pilot-in-command is *responsible for determining whether that aircraft is in condition for safe flight*. The PIC shall discontinue the flight when unairworthy mechanical, electrical, or structural conditions occur. Thus, the PIC must be as familiar as maintenance with the MEL and its procedures and especially the MEL applicable flight restrictions.

Based on the above, the obvious solutions to MEL problems are the following:

1. Air carrier training programs must provide additional and more complete training regarding the MEL for both maintenance and pilots.
2. Pilots must be as familiar as maintenance personnel with the MEL for the aircraft they are flying, *especially regarding flight restrictions*.

Additional discussion and evaluation are presented in Chap. 10.

3

Illegal and/or Unsafe Flight Due to Excessive Pressure for On-Time Departure

The airlines are constantly being criticized for unsatisfactory on-time performance, especially by the news media. This results in excessive pressure by management on both the maintenance personnel and the pilots.

Since the pilots have been given the ultimate responsibility for the airworthiness of their aircraft by the FAA, many times they find themselves in the situation called *pilot error* for accepting an aircraft against their better judgment due to excessive pressure for on-time departures.

The five ASRS reports included in this chapter present pilots' accounts of questionable maintenance procedures due to the emphasis on flight schedules. At the end of this chapter, a discussion and possible resolution of this problem area are presented. Additional discussion and evaluation are presented in Chap. 10.

MEL Not Complied with by Maintenance Due to Pressure for On-Time Departure (ASRS Report #156910)

Date: 1990/09
Type of operation: Air carrier, freight flight
Aircraft: Low-wing, heavy transport, over 300,000 lb
Engines: Four turbojets
Number of crew: Three
Flight crew: ASRS reporter: captain, pilot-in-command. 8000 hours, 150 hours within the past 90 days, 2800 hours in make/model. Air transport certificate.

Narrative (unedited, ASRS reporter's own words)

Fuel quantity gauges inoperative: Three—two in tanks not being used, one in a tank that was being used. Fuel tank that was being used was deferred under incorrect section of MEL (minimum equipment list) (thus it was not legal). If deferred under correct section, aircraft operation could have been conducted legally and safely. Further review of MEL was conducted inflight by crew. Found discrepancy and advised maintenance of MEL conflict.

Causes: Company's demand for on-time departure. Disciplinary actions taken against flight crews for failure to maintain schedule regardless of safety considerations. Pressures brought to bear on maintenance personnel as above indicated, lack of replacement equipment, and failure of company to provide adequate maintenance on equipment or purchase parts. Intimidation by company on personnel which cannot help but effect [sic] crew response to aircraft problems.

Report synopsis
Heavy transport aircraft was dispatched with inoperative fuel gauges. Minimum equipment list requirement was not complied with by maintenance dispatch.

Aircraft involvement
Aircraft was involved in an anomaly (an unsafe or illegal event).

Instances of Deferred Maintenance a Dangerous Policy (ASRS Report #116045)

Date:	1989/07
Type of operation:	Air carrier, passenger flight
Aircraft:	Low-wing, medium-large transport, between 60,001 and 150,000 lb
	Engines: Two turbojets
	Number of crew: Two
Flight crew:	ASRS reporter: captain, pilot-in-command. Air transport certificate.

Narrative (unedited, ASRS reporter's own words)
Experienced excessive pressure from maintenance supervisor to depart on schedule with mechanical

problems (due to no replacement parts available at this main base). When I delayed flight until both the dispatcher and I were satisfied, and I got a hard copy teletype authorization to proceed, I was required to report to my chief pilot to explain this delay. This happens constantly at this air carrier, and many pilots feel intimidated by the company whenever they take positions affecting safety.

Report synopsis
Reporter cited instances of deferred maintenance as a dangerous policy practiced by an air carrier.

Aircraft involvement
Aircraft was involved in an anomaly (an unsafe or illegal event).

Pressure for On-Time Performance Caused Maintenance Problems (ASRS Report #98676)

Date:	1988/11
Type of operation:	Commuter air carrier, passenger flight
Aircraft:	Low-wing, small transport, between 5001 and 14,500 lb
	Engines: Two turboprops
	Number of crew: Two
Flight crew:	ASRS reporter: captain, pilot-in-command. 4000 hours, 250 hours in the past 90 days, 1000 hours in make/model. Air transport certificate.

Narrative (unedited, ASRS reporter's own words)

Lately, there has been an increased emphasis on on-time performance. This has led to maintenance being put off, because it would lead to a delayed flight. In the following case, it directly led to the loss of instruments in flight. On a flight, I noticed there was over 10 degrees different [sic] between the captain's and F/O's (first officer's) DG (directional gyro). I wrote this discrepancy up in our Dallas hub. Because there was so little ground time, the maintenance department just changed the F/O's DG (HSI) (horizontal situation indicator) rather than really checking out the problem, which would have taken more time.

This did not correct the problem, and the following day I again wrote the problem up. Again, there was too little ground time for them to thoroughly check out the problem, so they only checked the system by swinging the compass. They did this because it was not time consuming, and then they signed it off as corrected. On the very next flight, after being airborne for about 15 minutes, the F/O's DG and the captain's RMI (remote magnetic indicator) compass card (they are slaved together) failed completely and started spinning constantly in a clockwise direction. Luckily, the weather was good, and I could proceed with the flight per MEL (minimum equipment list).

Trying to maximize profits, our company flies the airplanes virtually all day, every day with no time for maintenance. Also, there are not enough mechanics to perform the maintenance. Finally, the quest for perfect on-time performance prevents time being taken to perform needed maintenance. Unless there is some sort of regulatory intervention, I cannot foresee the problems decreasing.

Report synopsis
Flight crew complained that pressure for on-time performance caused maintenance problems.

Aircraft involvement
Aircraft was involved in an anomaly (an unsafe or illegal event).

Landing Gear Door Positioned Improperly (ASRS Report #87619)

Date: 1988/05
Type of operation: Air carrier, passenger flight
Aircraft: High-wing, medium-large transport, between 60,001 and 150,000 lb

Engines: Four turbojets

Number of crew: Two

Flight crew: ASRS reporter: first officer (F/O). 6000 hours, 100 hours within the past 90 days. Commercial license, instrument rating.

Narrative (unedited, ASRS reporter's own words)
Upon receiving aircraft for first flight of the day a normal preflight inspection was conducted by the F/O (first officer). F/O noticed maintenance was working on the R/H main landing gear brake fan. Passengers were boarded and maintenance advised myself that they were going to MEL (minimum equipment list) the brake

fan. The logbook was signed and aircraft released from maintenance. Everything was normal until the gear was selected up after takeoff. The R/H gear (red light) not in position selected, light remained on. The abnormal checklist was consulted and crew elected to return to the field.

Aircraft returned to the gate where maintenance inspected the R/H gear and found that they forgot to secure the gear door in the correct position. The mechanics involved were experienced people who were rushed to get the flight out on time. As the gear was retracted, but the gear door was not, aircraft performance was not significantly effected [sic]. However, it is the emphasis to get the flight out on time that may be detracting from the safety and thoroughness of maintenance procedures. To prevent this particular occurrence from happening again, I am determined to re-check maintenance's positioning of the gear doors whenever work has occurred on the tires or brakes.

Report synopsis
Landing gear door was positioned improperly by airline maintenance.

Aircraft involvement
Aircraft was involved in an anomaly (an unsafe or illegal event).

Takeoff after Warning Horn and without Maintenance Inspection (ASRS Report #135922)

Date/time: 1990/02

Type of operation: Air carrier, passenger flight

60 Chapter 3

Aircraft: Large, low-wing transport, between 150,001 and 300,000 lb
Engines: Three turbojets
Number of crew: Three
Retractable landing gear

Flight personnel: ASRS reporter: captain, pilot-in-command. 11,700 hours, 90 hours within the past 90 days, 7257 hours in make/model. Air transport certificate, flight engineer rating, instrument rating. Second officer (S/O). Flight engineer rating, commercial certificate, instrument rating.

Narrative (unedited, ASRS reporter's own words)

We were holding position on the runway for takeoff. When cleared, I started pushing throttles forward, and the takeoff warning horn sounded. I only had attained taxi speed, so we requested to taxi clear of the runway. Once clear, we checked that the flaps and leading edge devices were all extended, speed brake stowed and the horizontal stabilizer set for takeoff. Everything was verified in its proper setting. Next, I cycled the speed brake handle. Then I pushed up #3 throttle and again got the takeoff warning horn. The F/O (first officer) then moved the stabilizer trim from 5.8 units to 5.0 units and back to 5.8. This time, the takeoff warning horn did not sound when I advanced the #3 throttle. I assumed it was a stuck switch or relay that had been released by actuating the stabilizer trim.

We were already delayed one hour by ATC (air traffic control) for traffic congestion at our destination hub, and our passengers would miss their last connecting flights for the night if I delayed any further. I decided to return for takeoff instead of returning to the gate for further investigation by maintenance. Although the takeoff and flight were uneventful, my decision was wrong. I should have turned and let maintenance confirm the airworthiness of the aircraft. That will be my future procedure.

Report synopsis
Large transport has takeoff warning horn sound as throttles are advanced. Crew aborts takeoff, resets trim, and departs again without problem. Company policy dictated a return to gate for maintenance inspection.

Aircraft involvement
Aircraft was involved in an anomaly (an unsafe or illegal event).

Summary and Assessment: Illegal and/or Unsafe Flight Due to Excessive Pressure for On-Time Departure

Although these five ASRS reports have been singled out, pressure from management for on-time departures is an underlying factor in these as well as all ASRS reports. The pilot-in-command is required to determine the airworthiness of the aircraft per FAR part 91.7. This responsibility is compromised by excessive pressure, as well as by intimidation, from management for on-time departures.

What can we say about "pilot error" in cases where pilots accept an aircraft for flight against their better judgment in this environment? The obvious answer is to change the environment.

These ASRS reports to FAA/NASA are an attempt by the pilots to change the environment by alerting the regulatory agency to potential safety problems. However, based on the latest news media reports, the traveling public demands on-time departures. Safety is not even considered by the ill-informed public and the media. The fact is, pilots are going to have to live within this environment whether they like it or not. So, how can pilots cope with an operational environment they cannot change?

There is no easy answer to this question, but possibly pilots can prepare themselves by doing the following:

1. Pilots should thoroughly familiarize themselves with the aircraft's systems so they can intelligently discuss, and even question, maintenance regarding various procedures. Thus, pilots do not have to rely completely on accepting maintenance's assessment of the problem. Yes, pilots are subjected to a thorough training program in the aircraft's systems. Being pilots, though, they may not have been very attentive during a boring discussion of the aircraft's hydraulic system, for example. Also, many pilots may consider the training program to be complete, whereas it may not be able to include all eventualities. It is the opinion of this writer that most pilots don't do enough self-study. This self-study could include discussions with mechanics during nonpressure times between flights. A thorough understanding of the aircraft's systems can put pilots in a better position when justifying to management their refusal to accept an unsafe aircraft.

2. Also, as indicated in Chap. 4 regarding improper maintenance paperwork, the pilot must be thoroughly

Excessive Pressure for On-Time Departure 63

familiar with the maintenance paperwork such as the MEL and the logbook entries.

3. After maintenance, it behooves pilots to personally inspect the repair and ask pertinent questions. After all, the pilot-in-command is the final authority regarding the safety of the aircraft.

Additional discussion and evaluation are presented in Chap. 10.

4

Illegal and/or Unsafe Flight Due to Improper Maintenance Paperwork

Maintenance personnel are responsible for performing all maintenance operations and properly recording their work per FAR (Federal Aviation Regulations). However, it is the responsibility of the pilot-in-command to determine that all maintenance, repairs, and inspections have been accomplished by reviewing the aircraft logbook as well as other paperwork such as airworthiness and registration certificates.

Many airlines, especially commuter airlines, fly into airports without resident company maintenance personnel. Sometimes, the airline has a contract with the local FBO (fixed-base operator) for routine or simple maintenance. However, the FBO's general-aviation mechanic normally is not very familiar with a complex turbine-powered aircraft. Thus, proper maintenance, as well as the maintenance write-off, may be compromised.

It is important for the PICs to be familiar with the required maintenance checks and to be sure they are not flying an aircraft with an overdue check. Airline aircraft require A, B, C, and D checks within certain flight-time periods. General-aviation aircraft require 100-hour and annual inspections.

After the 13 ASRS pilot reports in this chapter, a brief summary and assessment will follow, but more comprehensive discussion and evaluation are presented in Chap. 10.

Flight Made without Properly Signed-Off Maintenance Item (ASRS Report #85180)

Date:	1988/04
Type of operation:	Air carrier, passenger flight
Aircraft:	Low-wing, large transport, between 150,001 and 300,000 lb
	Engines: Three turbojets
	Number of crew: Three
Flight crew:	ASRS reporter: captain, pilot-in-command. 15,000 hours, 210 hours within the past 90 days. Air transport certificate.

Narrative (unedited, ASRS reporter's own words)

On taxi out for takeoff, upper yaw damper failed to work to the right (OK to left and lower yaw damper OK). Although allowed to crew placard per MEL, I elected to return back to ramp to have maintenance look at it, and either fix it or placard the upper yaw damper and take the aircraft out of category II. They could not fix it and told me to take the aircraft and placard it in flight. This was my first error in that maintenance must sign off the problem at a major maintenance base, i.e., RDU (Raleigh Durham, NC). I checked with dispatch, and we agreed that the flight could be flown with the present fuel supply at the lower altitude and the longer flight time. We took the aircraft. I should not have allowed someone to talk me into taking an aircraft without the proper signoffs, even though we could have legally taken the aircraft if we had not returned back to the gate. Returning back to the gate changed the requirements, i.e., maintenance must correct the problem or restrict the aircraft, i.e., lower altitude, lower speeds, no category II, etc.

Report synopsis
Aircraft, a large transport flight, was made without properly signed off maintenance item.

Aircraft involvement
Aircraft was involved in an anomaly (an unsafe or illegal event).

Transport Flew Scheduled Passenger Flight without Proper Maintenance Checks Being Completed and Signed Off (ASRS Report #90444)

Date:	1988/07
Type of operation:	Air carrier, passenger flight
Aircraft:	Low-wing, medium-large transport, between 60,001 and 150,000 lb

	Engines: Two turbojets
	Number of crew: Two
Flight crew:	ASRS reporter: first officer (F/O). 3200 hours, 300 hours within the past 90 days. Air transport certificate.

Narrative (unedited, ASRS reporter's own words)

IAH (Bush Intercontinental Airport, Houston, TX). Upon arrival at aircraft, logbook was checked and maintenance called about a placard in regard to the LED (leading edge devices). The maintenance person advised us that we were OK to go after he reviewed the log and had verified that the flaps retracted (#2 LED) properly. We flew to PHL (Philadelphia, PA).

After arriving, it was brought to our attention that the aircraft may be overdue on a check. The mechanic called supervisor at LAX (Los Angeles) maintenance control to inform him that the log did not indicate that an "A" check was done. After their conversation, the mechanic told us that supervisor said it was OK to fly to DEN (Denver, CO). We were not going without a written message from maintenance. Maintenance and dispatch finally notified the flight was cancelled.

Upon further evaluation of the logbook, it appears that both the "A" check and service "C" checks were not done within the required amount of time prior to our flight from IAH-PHL. Cause: improper evaluation and record keeping by maintenance. Corrective action: better records.

Report synopsis

Aircraft, a medium-large transport, flew a scheduled passenger flight without proper maintenance checks being completed and signed off.

Aircraft involvement
Aircraft was involved in an anomaly (an unsafe or illegal event).

Plane Tagged as Airworthy by Maintenance When Beyond Annual Inspection (ASRS Report #112936)

Date: 1989/06
Type of operation: State government business flight
Aircraft: High-wing, small aircraft, less than 5000 lb, fixed gear
Engines: One reciprocating
Number of crew: One
Flight crew: Pilot-in-command. 1935 hours, 60 hours within the past 90 days, 425 hours in make/model. Commercial license, instrument rating, certified flight instructor.

Narrative (unedited, ASRS reporter's own words)

On Jun/Thu/89, I was assigned an aircraft for a flight from Springfield to Rockford for State of Illinois business to RON (remain overnight) and return to Springfield on Jun/Fri/89. A check of the maintenance clipboard and the schedule board showed the aircraft to be "green tagged" as airworthy for flight from a maintenance standpoint.

The flight was routine, other than enroute [sic] IFR (instrument flight rules) flight plan was necessitated due to IMC (instrument meteorological conditions) that did not improve as forecast. I did not become aware of the fact that the aircraft was out of annual inspection until the morning of Jun/XX/89 when I received a call from our maintenance department informing me of a "mix up." I obtained a ferry permit and an A&P (aircraft and powerplant mechanic) check of the aircraft and conducted a routine VFR (visual flight rules) flight back to Springfield the evening of Jun/XX/89.

I realize that determining the airworthiness of an aircraft is the responsibility of the PIC (pilot-in-command). In the future I will personally check maintenance records on aircraft I am assigned to fly and not put "blind faith" in a maintenance department.

Report synopsis
The pilot flew the aircraft after checking the maintenance clipboard and noting the aircraft "green-tagged" as airworthy. After landing elsewhere, the pilot was notified by phone that the aircraft was beyond the annual check. The pilot obtained a ferry permit and had A&P check and flew home.

Aircraft involvement
Aircraft was involved in an anomaly (an unsafe or illegal event).

Flight Took Off Again without Obtaining Rerelease from Dispatcher after Return for Pressurization Maintenance (ASRS Report #135419)

Date: 1990/01
Type of operation: Air carrier, passenger flight

Aircraft:	Low-wing, medium-large transport, between 60,001 and 150,000 lb
	Engines: Two turbojets
	Number of crew: Two
Flight crew:	ASRS reporter: captain, pilot-in-command. 12,500 hours, 180 hours within the past 90 days, 6500 hours in make/model. Air transport certificate, flight engineer rating.

Narrative (unedited, ASRS reporter's own words)

After takeoff from ROC (Rochester, NY) aircraft would not pressurize. It was obvious that all air conditioning and pressurization systems were operating normally and that a "hole" existed in the airplane. I elected for an immediate return to ROC. We found door 2R was closed in a cocked position. The cams on the leading edge of the door had not engaged their proper slots. The door warning light was not on in spite of this condition. Had local mechanic inspect door for damage, etc., and clear log with concurrence of our maintenance folks in SFO (San Francisco, CA).

As we had a dual fuel load to start and still have 4.5 over minimum load for the leg, we took off again about one hour late. However, I failed to contact dispatcher personally for a re-release, which made dispatch unhappy and rightfully so. I was busily involved in clearing the log and just assumed that dispatch was informed by others in the company. A complicating factor was that our company radio failed to report our landing and block times back in ROC; thus dispatch was not aware of the event until after we were airborne the second time. This is a good reminder that after any event, the first person to contact is dispatch because of their joint responsibility for the flight.

Report synopsis
Flight crew of medium-large transport was unable to pressurize aircraft after takeoff. They returned and landed. Company radio did not advise dispatch. Flight took off after maintenance, failed to obtain rerelease, which was in noncompliance with company and FAA policy.

Aircraft involvement
Aircraft was involved in an anomaly (an unsafe or illegal event).

FAA Inspector Files Violations for Expired Registration and Lack of Weight and Balance Computation (ASRS Report #114431)

Date:	1989/06
Type of operation:	Corporate aircraft, light cargo flight
Aircraft:	Low-wing, light general-aviation transport, between 14,501 and 30,000 lb
	Engines: Two turbojets
	Number of crew: Two
Flight crew:	ASRS reporter: captain, pilot-in-command, general-aviation pilot. 8200 hours, 50 hours

Improper Maintenance Paperwork

within the past 90 days,
600 hours in make/model.
Air transport certificate.

Narrative (unedited, ASRS reporter's own words)

Basically, I am writing this report for the purpose of immunity, as I am not convinced that there is a safety problem from which we could all learn. After takeoff, my landing gear would not retract, so I returned to the field and landed. Shortly after landing, the FAA maintenance inspector showed up and began looking for things wrong with the operation. By the way, keep in mind that I was flying an airplane which I have been flying for the same company for over 4 years without an interruption.

Unknown to me, the company changed the registered name and there was a temporary registration slip in the aircraft. That temporary registration had expired the previous day! Therefore, the maintenance inspector knew he had a bonafide [sic] criminal here and proceeded to stress my culpability in the crime of flying an illegal airplane. Next, the inspector checked our weight and balance. My next violation came to light.

Prior to receiving our cargo of AOG aircraft parts, the company dispatchers had asked our on-duty mechanic to remove interior seats to avoid soiling or damage. The inspector looked at our form for showing the pilots the weight and balance procedures and noticed that it made no provisions for the seats being out. When asked, I said that the aircraft was lighter, if anything, and that I knew that the aircraft was well within the center of gravity and weight limits. He was unsatisfied and [said] that another violation could be filed against me. After our company chief pilot showed up, he had me work out a weight and balance longhand (our system used a template) and that computation showed that we were 2500 lbs. under gross and with the center of gravity within limits.

When I left, the maintenance inspector was also making noises about violating the company for not providing accurate information to its pilots (center of gravity with seats out) and was loudly criticizing the mechanic, because he raised the airplane and cycled the gear without following the manufacturer's troubleshooting guide. This guy came on like "McGarret, Five-Oh," and seemed to enjoy it. This new attitude that the FAA has of "making them pay" is doing nothing for the safety of the traveling public, endangering the livelihood of professionals, and causing pilots to go underground with problems out of fear. Such policies will have that exact opposite effect of that which the public desires in the long run.

Report synopsis
FAA maintenance inspector on ramp inspection discovered that the temporary registration was one day past expiration. Then he challenged removal of seats without rework of the weight and balance.

Aircraft involvement
Aircraft was involved in an anomaly (an unsafe or illegal event).

Owner-Pilot Did Some of His Own Maintenance, but Failed to Have Mechanic Sign Logbook (ASRS Report #122749)

Date: 1989/09
Type of operation: General-aviation flight, not for compensation

Improper Maintenance Paperwork

Aircraft: High-wing, fixed-gear, small aircraft, less than 5000 lb
Engines: One reciprocating
Number of crew: One

Flight crew: ASRS reporter: owner-pilot. 680 hours, 90 hours within the past 90 days, 75 hours in make/model. Commercial license with instrument rating, certified flight instructor.

Narrative (unedited, ASRS reporter's own words)

One week ago, I asked for and received maintenance services from maintenance operations, MGW (Morgantown, WV). At that time, I indicated to the mechanic that I needed, and would personally replace, several load mounts on the aircraft.

The requested work was completed; however, the mechanic also left a note indicating that the aircraft was unairworthy until the receptacles were replaced. I replaced the parts with approved parts and believed, under preventive maintenance, that I was authorized to do so. On 9/X/89, I was pilot-in-command (PIC) and made one landing. Following the landing (with one passenger), I became aware that the throttle did not engage properly, and I pulled off the ramp. I removed the cowling and discovered that the throttle was not properly attached (no nut, washer, or safety pin). The work was supposedly completed the past week by an authorized mechanic at the FBO (fixed base operator).

I brought the matter to the attention of the FBO manager. The manager indicated that the mechanic could have overlooked the item. He also indicated that I was

78 Chapter 4

flying an unairworthy aircraft until the mechanic signed the aircraft logbook, indicating that the replaced load mounts were acceptable. I indicated that I would make arrangements to have the mechanic inspect the work for return to service.

Report synopsis
The owner-pilot chose to do some of his own maintenance. Aircraft was not airworthy until signed by A&P (aircraft and engine mechanic). The pilot flew the aircraft prior to the entry being made. The pilot discovered a maintenance error as well.

Aircraft involvement
Aircraft was involved in an anomaly (an unsafe or illegal event).

Aircraft Airworthiness, Registration, and Maintenance Records Not in Order (ASRS Report #137871)

Date:	1990/02
Type of operation:	Test flight of rental small aircraft
Aircraft:	High-wing, small aircraft, less than 5000 lb, retractable gear
	Engines: One reciprocating
	Number of crew: Single pilot undergoing instruction.
Flight crew:	General-aviation pilot. 265 hours, 15 hours within the past 90 days, 0 hours in make/model. Commercial license, instrument rating.

Narrative (unedited, ASRS reporter's own words)

On Feb/Wed/90, I was taking my check ride for the certified flight instructor license. I took the test through the FAA by way of the local FSDO (flight service district office) located at the airport. The aircraft I used was rented from XXXX, also located at Allegheny County Airport.

A complex aircraft was hangared and had just undergone gear work by a certified mechanic because of gear problems in earlier flights. This was brought to my attention, and I replied that I would use the aircraft for my test if I could fly it first to familiarize myself with the aircraft.

Since the aircraft needed to be test flown, I was to be checked out in it and the aircraft test flown simultaneously. I and a flight instructor from XXXX were the pilots. After the flight, the aircraft's logbooks were taken to the FAA for approval of airworthiness as required by the FAA for all of their check rides.

Shortly thereafter, I was notified that the documents on board the aircraft were invalid and the aircraft was flown illegally by: 1) The aircraft had on board a temporary registration certificate which had expired. 2) The N number on the airworthiness certificate was different from the N number on the aircraft. The serial number on the airworthiness certificate did not agree with aircraft registration and log books. 3) No logbook entry in the aircraft logbooks were [*sic*] made for the gear work and the test flight sign-off.

Aircraft N number: XA. I was PIC (pilot-in-command) and by FAR part 91.3, I was responsible for that flight and I should have checked the aircraft documents. I have flown aircraft at flight schools all of my flying career. The aircraft at flight schools are numerous, and it is difficult to know all the details of every aircraft. To prevent further problems of this nature in the future, aviation flight schools need to be more careful in aircraft documents.

Report synopsis
Aircraft airworthiness, registration, and maintenance records were not in order. This resulted in illegal flight.

Aircraft involvement
Aircraft was involved in an anomaly (an unsafe or illegal event).

Flight Conducted with Unresolved Maintenance Items (ASRS Report #137893)

Date:	1990/02
Type of operation:	Air carrier, passenger flight
Aircraft:	Low-wing, medium-large transport, between 60,001 and 150,000 lb
	Engines: Two turbojets
	Number of crew: Two
Flight crew:	ASRS reporter: captain, pilot-in-command. 12,300 hours, 220 hours within the past 90 days. Air transport certificate, flight engineer rating.

Narrative (unedited, ASRS reporter's own words)
Flying aircraft with 2 open items in logbook. The above were discovered by FAA on arrival at a maintenance

station. Previous crew that wrote up the items apparently reset the electrical system cross-tie. The cross-tie lockout light was not on when we arrived at aircraft. I noticed #3 fuel tank circuit breaker out and reset it. It never popped again and pump operated normally.

When I first arrived at aircraft I asked the copilot if anything was in the log. He said no. I found out later his normal procedure was to check only the front of the logbook at non-maintenance stations, since it was the inbound crew's job to notify technical service of any write-ups.

Since the above problem could happen again, I feel it should be stressed that: 1) Both crew members should check the log, not only the current incident section but the last several pages also. 2) It is even more important at a non-maintenance station, since there are no maintenance personnel to check the book and the aircraft usually arrives at non-maintenance station late at night when the previous crew is tired (more open to error).

Report synopsis
Air carrier flight was conducted with unresolved maintenance items.

Aircraft involvement
Aircraft was involved in an anomaly (an unsafe or illegal event).

Captain Coerced into Accepting an Aircraft with Door Latch Problems (ASRS Report #156013)

Date: 1990/09
Type of operation: Air carrier, passenger flight

Chapter 4

Aircraft:	High-wing, medium transport, between 30,001 and 60,000 lb
	Engines: Four turboprops
	Number of crew: Two
Flight crew:	ASRS reporter: captain, pilot-in-command. 8000 hours, 215 hours within the past 90 days, 3000 hours in make/model. Air transport certificate, flight engineer rating.

Narrative (unedited, ASRS reporter's own words)

While waiting for aircraft to come in from PHL (Philadelphia, PA), talked with dispatch concerning aircraft mechanical problems (we were late and another crew was bringing it in from PHL). I was told that the problem was with the door seal (pressurized aircraft). When the aircraft arrived in DCA (Ronald Reagan National, Washington, DC), I talked with the crew and found out that it was the door closing mechanism. A lever is pushed down over center, which inflates the door seal and keeps the door immobile. (The door is heavy, and this same lever, when lifted, raises the whole door over lips so that it can open outward).

The other crew told me that the door was fixed with no problems on their flight but that the lever was "spongy." It was, but all door lights went out when it was closed, so I accepted the aircraft. Then I discovered that the pressurization had been MEL'd (minimum equipment list). Although the door seal (pressurization) is an integral part of the lever, the lever itself was what had been giving problems.

Improper Maintenance Paperwork 83

I took the aircraft as previous crew had said "no problem." Enroute [sic], the handle popped up just over center, and the door seal deflated. A passenger asked the F/A (flight attendant) if she needed help holding it down (she said no and left the lever as it was. Door did not open).

My main concern is that in my opinion (after the fact), I feel that the aircraft should not have been in revenue service. I also feel that the company had a lot more facts than I did. They maneuvered me and my crew into operating a questionably airworthy aircraft.

Report synopsis
Captain of air carrier, medium-transport complained about being coerced into accepting an aircraft with a maintenance listing of door latch problems. The door latch malfunctioned in flight.

Aircraft involvement
Aircraft was involved in an anomaly (an unsafe or illegal event).

Flight Crews Failed to Note That Logbook Lacked the Required Maintenance Release Sign-Off (ASRS Report #157254)

Date:	1990/09
Type of operation:	Air carrier, passenger flight
Aircraft:	Low-wing, medium-large transport, between 60,001 and 150,000 lb
	Engines: Two turbojets
	Number of crew: Two

84 Chapter 4

Flight crew: ASRS reporter: captain, pilot-in-command. 9600 hours, 200 hours within the past 90 days, 6400 hours in make/model. Air transport certificate.

Narrative (unedited, ASRS reporter's own words)

Earlier [in the] week, company aircraft #AB had sustained hail damage. Aircraft damage had been repaired and noted as back in service in logbook. Several days later, I flew same aircraft and noticed no logbook discrepancies. A call from another pilot indicated that our company manual required a new airworthiness release (in addition to the clearing of hail damage written) be in the forms. I believed that simply clearing the write-ups was sufficient and that the airworthiness release would only have been necessary if the maintenance folks had found no damage after inspection. Did not intentionally fly without a new airworthiness release.

Supplemental information: On Friday 9/90M at AM:50 local, I, as captain, accepted an aircraft for a series of flights. The preflight was normal with no discrepancies. The previous evening prior to reaching the out station for the overnight, the previous captain wrote a discrepancy at a maintenance base to "inspect radome for possible hail damage." The corrective action was "R and R radome, operations checked, OK per maintenance manual." Aircraft on a series of flights. According to our flight operations manual, with suspected hail damage, an airworthiness release should have been accomplished at the maintenance base. I did not catch the error.

Additional supplemental information: I learned a few days later that this write-up requires an airwor-

Improper Maintenance Paperwork 85

thiness release according to the operations manual, upon accepting the aircraft instead of trusting to maintenance to correctly sign off the write up. Maintenance did sign off an airworthiness release on September Friday in the afternoon.

Report synopsis
A series of flights made by different flight crews on the same aircraft failed to note that the logbook did not carry the required maintenance release sign-off after radome replacement.

Aircraft involvement
Aircraft was involved in an anomaly (an unsafe or illegal event).

Aircraft Put Back in Operation with Repairs Completed but Not All Logbook Items Cleared (ASRS Report #175770)

Date:	1991/04
Type of operation:	Air carrier, passenger flight
Aircraft:	High-wing, medium transport, between 30,001 and 60,000 lb
	Engines: Four turboprops
	Number of crew: Two
Flight crew:	ASRS reporter: captain, pilot-in-command. 4600 hours, 210 hours within the past 90 days, 1300 hours in make/model. Air transport certificate.

Narrative (unedited, ASRS reporter's own words)

Upon taxiing into the ramp at GJT (Grand Junction, CO) we shut down engines 2 and 3. We immediately lost the right DC electrical bus (bus tie split) including the control of power nosewheel steering and interphone communications between captain and F/O (first officer). I stopped the aircraft, could not fix the problem and continued to taxi to the parking spot.

As soon as DC external power was introduced, the #3 engine starter/generator engaged. The problems were written in the logbook, and the airplane was grounded for several hours while company sent a mechanic with parts to fix it. After fixing the original problem, 2 more problems surfaced. The aircraft now would not accept external power, and the #1 DC generator was not on line [*sic*].

These new problems were written up in the logbook. A couple more hours passed while the company sent another mechanic with parts for the new problems. When he arrived, he and the first mechanic immediately fixed the DC external problem while I watched (it took only 1–2 minutes), then they began working on the final problem which they could not fix. So the #1 DC generator was deferred with the appropriate logbook entry (the original problem had also been cleared in the logbook).

Afterwards, we flew the aircraft back to our maintenance base in Denver. However, due to the successive similar problems that occurred, the time that elapsed (7 hours total on the ground in GJT), the confusion from 2 separate mechanics being sent out, etc., and lastly, the simple cure of the DC external power problem, at no time did the DC external power problem get cleared in the logbook. As PIC (pilot-in-command) I should have caught the mistake in the logbook before we took off. Also contributing to my laxness was that we carried no passengers on the flight from GJT to DEN.

Report synopsis
After 7 hours of maintenance activity, the medium transport aircraft was put back in operation without all the logbook items being cleared. Repairs were completed, but the paperwork was not completed.

Aircraft involvement
Aircraft was involved in an anomaly (an unsafe or illegal event).

Open Item in Maintenance Log (ASRS Report #295085)

Date:	1995/01
Type of operation:	Air carrier, passenger flight
Aircraft:	Brasilia EMB-120, low-wing, medium transport, between 30,001 and 60,000 lb
	Engines: Two turboprops
	Number of crew: Two
Flight crew:	ASRS reporter: captain, pilot-in-command. 9000 hours, 245 hours within the past 90 days, 6000 hours in make/model. Air transport certificate.

Narrative (unedited, ASRS reporter's own words)
We traded aircraft with another crew. I was not informed that the aircraft had an open discrepancy written up in

Chapter 4

logbook. I flew 3 legs (SFO to RNO) (San Francisco to Reno) in said aircraft. Apparently our maintenance had looked at this aircraft and could not find any problems but had failed to clear the write up [sic]. I checked the logbook prior to departing, but I did not notice the write up had not been cleared.

Report synopsis
There was an open item in maintenance log.

Aircraft involvement
Aircraft was involved in an anomaly (an unsafe or illegal event).

Small Transport Grounded for Several Discrepancies, and FAA Held Last Pilot in Noncompliance (ASRS Report #427449)

Date:	1999/01
Type of operation:	Air taxi, cargo flight
Aircraft:	Low-wing, small transport, between 5001 and 14,500 lb
	Engines: Two reciprocating
	Number of crew: One
Flight crew:	ASRS reporter: single pilot, pilot-in-command. 15,500 hours, 200 hours within the past 90 days, 800 hours in make/model. Air transport certificate, flight instructor.

Improper Maintenance Paperwork

Narrative (unedited, ASRS reporter's own words)

During an FAA inspection of our air taxi a number of discrepancies were found by the FAA inspector on a C402 (Cessna 402) that I had last flown the previous day. Since I was last to have flown the aircraft, it was alleged that I had flown an unairworthy aircraft. The discrepancies include: 1) A leaking hydraulic ram under the left landing gear, 2) A crushed bulkhead and deformed skin around the tail tie down ring/loops, 3) A bent crossbrace in the nose gear well, 4) Red residue on top of the left engine nacelle.

From the outset let me make a statement that my practice is to do a thorough preflight prior to each flight. If I had knowledge of the discrepancies, I would have grounded the aircraft in question and sought out corrective measures to remedy the situation. Explanations: We have been experiencing extremely low temperatures (-40 degrees C) here in Nome, AK. The C402 was sitting on the ground out in the cold and was pulled into the hangar where the leak was discovered. The leaking apparently started after the airplane came into the hangar. Our ramp is asphalt covered by snow that has been graded for removal. After the leak was noted, I checked outside on the ramp and saw no red drops or puddles of hydraulic fluid. This seems to confirm that the leak started after it was in the hangar and was most likely due to the extreme cold temperatures.

The aircraft is occasionally loaded with cargo inside the hangar due to the extreme cold and dark conditions out on the ramp. The loading is generally done by the cargo personnel. The apparent crushing damage in the tail could have taken place if the cargo handler had loaded up the airplane tail heavy and while pushing it out or pulling it into the hangar the tail hit the ground, since the ramp is both steep and slippery due to the covering of snow and ice. The airplane was loaded for a flight the day of the inspection but was then unloaded to facilitate the base inspection's requirements. The damage done did not happen from a hard landing, because there have not been any, and the pattern of the damage does not appear to be the result of a scraping nature.

The bent cross tube in the nose gear well appears to be something new since the metal is shiny where the damage occurred. Nome, Alaska is right on the coast of the Bering Sea, and the salt air tends to tarnish and discolor any exposed fresh metal scratches. The damage appears to have resulted from the nose gear not fully extending after takeoff prior to being retracted into the gear well. This was probably due to the extreme temperatures where the nitrogen volume contracts and can leak out with the hydraulic fluid. Red residue—I am not sure where it came from. It should be noted that there is a hydraulic pump on the accessory case of the left engine.

Callback conversation with the reporter revealed the following information: Reporter stated that the FAA is not going to take pilot certificate action, as it believes that the discrepancies were more the responsibility of the company's maintenance procedures together with the cargo handlers.

Report synopsis

FAA inspection of an air taxi cargo operator disclosed that a C402C Utiliner had discrepancies which grounded the aircraft. The FAA held the last pilot to fly the aircraft, the reporter, in noncompliance for flying an unairworthy aircraft.

Aircraft involvement
Aircraft was involved in an anomaly (an unsafe or illegal event).

Summary and Assessment: Illegal and/or Unsafe Flight Due to Improper Maintenance Paperwork

The following is a summary of the ASRS reports in this chapter:

The pilot-in-command accepted an aircraft for flight without proper maintenance sign-off. Maintenance should have corrected the yaw damper problem or imposed flight restrictions (ASRS report #85180).

The PIC flew an air carrier aircraft with overdue A and C service checks and did not notice write-ups (ASRS report #90444).

In ASRS report #112936, a business aircraft pilot flew small reciprocating engine aircraft cross-country with overdue annual inspection. Pilot did not personally check the maintenance records.

Flight crew was unable to pressurize aircraft after take-off and returned to gate. Flight took off again after maintenance, but the flight crew failed to obtain maintenance re-release (ASRS report #135419).

After landing a light general-aviation transport, the pilot was met by an FAA maintenance inspector. The aircraft's temporary registration was one day expired. Then the pilot was told that removal of seats required weight and balance rework (ASRS report #114431).

In ASRS report #122749, an owner-pilot chose to do some of his own maintenance. Aircraft was not airworthy until signed off by A&P (airframe and power plant mechanic). Pilot flew aircraft prior to maintenance sign-off.

General-aviation aircraft airworthiness, registration, and maintenance records were not in order, resulting in illegal flight (ASRS report #137871).

In ASRS report #137893, the captain of a light transport asked the copilot to check the log for maintenance items. The copilot indicated no maintenance problems but had checked only the front of the logbook and missed unresolved maintenance items on the last several pages. The air carrier flight was conducted with unresolved maintenance items.

The captain of an air carrier aircraft with door latch problems accepted the aircraft for flight against his better judgment. The door malfunctioned in flight. The aircraft had maintenance listings of unresolved door latch problems (ASRS report #156013).

In ASRS report #157254, a series of flights made by different flight crews on the same aircraft failed to note that the logbook did not carry the required maintenance sign-off after radome replacement.

In ASRS report #175770, the PIC flew air carrier aircraft with repairs not written off.

The PIC of air carrier aircraft did not notice that the logbook write-up had not been cleared and flew three legs in the aircraft (ASRS report #295085).

In ASRS report #427449, FAA inspection of an air taxi aircraft disclosed that the C402C Utiliner had discrepancies which grounded the aircraft. The FAA held the last pilot to fly the plane, the reporter, in noncompliance for flying an unairworthy aircraft.

In all the ASRS reports in this chapter, the pilot-in-command did not discover open items in the maintenance paperwork; therefore, the PIC accepted an illegal and/or unsafe aircraft for flight.

Although many errors in maintenance write-ups were those of maintenance personnel, it is the PIC who is ultimately responsible for accepting a legal and airworthy aircraft.

Both the pilot and the copilot should check the logbook, not only the current incident section, but the last several pages also. This is even more important at a nonmaintenance station since there are no company maintenance personnel to check the log, and the aircraft may arrive late at night when the previous crew is tired and more susceptible to error.

In other words, pilots must pay strict attention to the logbook and all maintenance paperwork. After all, the PIC is ultimately responsible for flying an airworthy aircraft.

Additional discussion is presented in Chap. 10.

5

Missed Checklist or Preflight Items

A turbine-powered aircraft is a complex machine incorporating many systems. All systems must be in airworthy condition for safe flight. Also, some systems are in different configurations for different flight phases, such as landing gear and flaps systems.

Detailed checklists have been devised so that the pilots can ascertain that the aircraft is configured properly for each phase of the flight.

In accordance with FAR part 91.503, for large and turbine-powered multiengine airplanes, each cockpit checklist must spell out the following activities and shall be used by the flight crew members when operating the airplane:

1. Before starting engines
2. Before takeoff
3. Cruise
4. Before landing
5. After landing

6. Stopping engines
7. Emergencies

Each emergency cockpit checklist must cover the following procedures, as appropriate:

1. Emergency operation of fuel, hydraulic, electrical, and mechanical systems
2. Emergency operation of instruments and controls
3. Engine-inoperative procedures
4. Any other procedures necessary for safety

Before flight, an exterior preflight inspection must be conducted, preferably by the pilot-in-command, or the flight engineer (second officer) of a three-crewmember aircraft. This inspection includes, but is not limited to, the following:

1. Visually check the entire airplane for obvious structural damage.
2. Ensure that all access doors are secure.
3. Verify that landing gear downlock pins have been removed.
4. Check that the control surface locks have been removed.
5. Ascertain that there are no fuel or oil leaks.
6. Make sure that engine air ducts are clear of foreign objects.
7. Verify that tires are properly inflated and in acceptable condition.
8. Check that fuel, oil, and other liquid systems have been properly serviced.

The following ASRS reports are examples of missed checklist incidents which resulted in unsafe flight.

A summary and an assessment of these ASRS reports are presented at the end of this chapter. Chapter 10 presents an overall assessment and lessons learned from this chapter's ASRS reports, as well as those of Chaps. 2 through 9.

Tail Damage Found Only after Several Legs Following Possible Tail Strike and Maintenance Inspection and Clearance (ASRS Report #114479)

Date: 1989/06
Type of operation: Air carrier, passenger flight
Aircraft: Low-wing, large transport, between 150,001 and 300,000 lb

Engines: Three turbojets

Number of crew: Three

Flight crew: ASRS reporter: captain, pilot-in-command. 9000 hours, 180 hours within the past 90 days, 600 hours in make/model. Air transport certificate.

Narrative (unedited, ASRS reporter's own words)

After powerback [sic] incident at DFW (Dallas/Ft. Worth, TX) and subsequent inspection and OK by maintenance, aircraft flew from DFW to MEX (Mexico City). Aircraft then flew from MEX to ORD (O'Hare, Chicago, IL). Five hours later, damage was found to tail of aircraft. Informed by chief pilot's office we had flown a damaged aircraft from MEX to ORD. Chief pilot told me that I and the engineer are both ripe for FAA action.

100 Chapter 5

Maintenance has cherry picker, spotlights, lots of people. They don't find damage until 5 flights later, in hangar, using hardstands. Flight crew was supposed to see it at 2 AM in Mexico City with his flashlight. Ask me if my mind is on flying the next leg or on this BS. Eighteen years and I've never scratched an airplane, and now the chief pilot says I'm going to have an FAA hearing.

Report synopsis
After power back from gate, flight crew was notified of possible tail strike with another aircraft. Maintenance inspected and determined negative damage. Several legs were completed and tail damage was found later.

Aircraft involvement
Aircraft was involved in an anomaly (an unsafe or illegal event).

Landing Gear Would Not Retract on Takeoff (ASRS Report #117625)

Date:	1989/07
Type of operation:	Air carrier, passenger flight
Aircraft:	Low-wing, wide-body transport over 300,000 lb
	Engines: Three turbojets
	Number of crew: Three
Flight crew:	ASRS reporter: second officer, flight engineer (S/O, F/E). 5000 hours, 210 hours within the past 90 days,

250 hours in make/model.
Air transport certificate, flight engineer rating.

Narrative (unedited, ASRS reporter's own words)

Event: A landing gear downlock pin was left in the center landing gear of an airplane on a scheduled flight DFW (Dallas-Ft. Worth, TX) to ORD (O'Hare, Chicago, IL).
Chain of events: Problem discovered after takeoff from DFW when landing gear were retracted. The center gear failed to come up. Landing gear cycled up, then down, but this failed to correct problem. DFW tower relayed through departure control that center gear was in down position. **Corrective actions:** Aircraft diverted to TUL (Tulsa, OK) to have gear checked. Landing at TUL uneventful. Maintenance discovered downlock pin in center gear. Pin removed. Aircraft dispatched to ORD.

How problem arose: Aircraft had been at DFW maintenance hangar for previous 2 days. Gear pin thought to be installed at hangar. Maintenance failed to remove pin after towing aircraft from hangar to departure gate. Mechanic assigned to help ensure on-time departure failed to discover gear pin in his preflight inspection. F/E (flight engineer) did not discover gear pin on his preflight inspection. Push crew at gate did not discover gear pin. It is the opinion of this writer that the gear pin did not have a flag attached that would have made the pin easy to see. With a flag attached it is hard to believe that the gear pin would not have been seen by someone who was preflighting the aircraft and as such was checking to make sure all gear pins were removed before flight. **How to prevent in future:** 1) All gear pins must have flags attached to them to make them obvious. 2) Stress to all concerned (maintenance, tow crews, etc.) the absolute importance of ensuring the downlock pins are [not] left in place.

Supplemental information: Event: F/E did not personally conduct exterior preflighting inspection. **Chain of events:** Diverted to TUL (DFW-ORD) because gear downlock pin not removed from center landing gear. Maintenance conducted exterior walk-around after removing gear pin. (**Note:** Separate NASA report submitted on gear pin). How problem arose from a change from the routine, i.e., divert, extra company communications, having to change destination after takeoff, etc. Human factors: 1) Crew upset at having to divert. 2) Crew members worried about who is to blame for gear pin left in place. 3) What will be punitive actions of company and FAA for having to divert? 4) Change in routine. 5) Being in the "spotlight"—there were a lot of men in "coats and ties" observing from ramp level at gate TUL.

How to prevent in future: 1) Follow established procedures. 2) Do not rush or get in a hurry. 3) From FAA and company standpoint, do not make crews feel as though their little mistake will cost them their jobs. This increases pressure on the crew members involved and compounds problems.

Report synopsis
Air carrier aircraft diverted to another airport where maintenance can check gear which would not retract on takeoff.

Aircraft involvement
Aircraft was involved in an anomaly (an unsafe or illegal event).

Engine Flameout and Emergency Landing Due to Reliance on Only a Verbal Assurance of Fuel Quantity (ASRS Report # 126630)

Date: 1989/10
Type of operation: Corporate aircraft, test flight after maintenance
Aircraft: Low-wing, light transport, between 14,501 and 30,001 lb
Engines: Two turboprops
Number of crew: Two
Flight crew: ASRS reporter: general-aviation pilot, captain, pilot-in-command. 11,398 hours, 103 hours within the past 90 days, 279 hours in make/model. Air transport certificate, certified flight instructor.

Narrative (unedited, ASRS reporter's own words)

Following repair work at John Wayne airport (SNA) spanning 17 days involving fuel tank leaks and bleed air problems, I accepted aircraft for a local test flight. We were told by mechanic in charge of this aircraft that it was fueled to 1200 lbs. per wing and ready for flight. We were not given a fuel delivery receipt.

After completing a preflight inspection with copilot/mechanic, I started the engines and taxied to the takeoff position. "Fuel transfer" warning lights were on during taxi but extinguished shortly following boost pump activation. During takeoff all indications were normal. The right "fuel transfer" light illuminated prior to landing and the fuel gauges indicated zero intermittently. The right fuel transfer light extinguished following right auxiliary boost pump activation. I suspected faulty cannon plug connections or maintenance oversight. I landed normally at Corona airport, parked the aircraft, shut down the engines and exchanged trainee pilots.

I departed Corona airport en route John Wayne airport, and the right engine flamed out approximately 3 minutes after takeoff. Relight procedures were unsuccessful, and I feathered/secured it. Approximately 90 seconds later, the left engine flamed out. Relight procedures unsuccessful. I feathered it and sought a suitable landing site. Accomplished a successful off-airport emergency landing (no injuries; minimal aircraft damage). The maintenance performed was not entered in the aircraft logbook, nor was I given a "maintenance release" document. FAA inspectors at the scene said this would be an incident, not an accident, and NTSB need not be notified.

In the future, I will visually check the fuel quantity or obtain an authenticated fuel delivery document or witness the fueling. I will not operate an aircraft with any abnormal indications, momentary or otherwise, regardless of expediency.

Report synopsis

Corporate light transport flight crew took the aircraft out on a test flight after maintenance and being told they had sufficient fuel. Their failure to visually check their fuel quantity resulted in flameout of both engines and an emergency off-airport landing.

Missed Checklist or Preflight Items 105

Aircraft involvement
Aircraft was involved in an anomaly (an unsafe or illegal event).

Small Transport Descended below Assigned Altitude While on Autopilot (ASRS Report #133799)

Date:	1990/01
Type of operation:	Test flight of corporate aircraft
Aircraft:	Low-wing, small transport, between 5001 and 14,500 lb
	Engines: Two turboprops
	Number of crew: One
Flight crew:	ASRS reporter: single pilot, pilot-in-command. 2100 hours, 35 hours within the past 90 days, 30 hours in make/model.

Narrative (unedited, ASRS reporter's own words)
Approximately 15 NM (nautical miles) out from Olympia airport, while being vectored for an approach, approach control gave us an altitude alert. Our assigned altitude was 2000 ft. Aircraft descended to approximately 1400 ft. while on autopilot with altitude hold engaged. Pilot's attention was outside the aircraft looking for airport and traffic. After regaining assigned altitude, it was discovered the trim disconnect was in the "off" position, and the autopilot would not hold altitude even though it would couple and capture the assigned altitude initially.

The aircraft had come out of maintenance, and pilot failed to use the full checklist and instead relied on an abbreviated checklist, normally used for quick turnarounds. The autopilot will accept altitude hold with the trim turned off. Also the pilot was in error for not using the full checklist after maintenance.

Report synopsis
The pilot of a small transport checking recent maintenance items descended below the assigned altitude while on autopilot. Pilot used abbreviated checklist instead of full checklist after maintenance.

Aircraft involvement
Aircraft was involved in an anomaly (an unsafe or illegal event).

Flight Crew Raises Gear on a Gear-Down Ferry Flight (ASRS Report #297664)

Date:	1995/03
Type of operation:	Ferry flight
Aircraft:	Beech 1900, low-wing, light transport, between 14,501 and 30,000 lb
	Engines: Two turboprops
	Number of crew: Two
Flight crew:	ASRS reporter: captain, pilot-in-command. 14,100 hours, 200 hours within the past 90 days, 3500 hours in make/model. Air transport certificate.

Narrative (unedited, ASRS reporter's own words)

On the inbound flight, when the landing gear was selected down, absolutely nothing happened. We cycled it once. It extended normally. We proceeded with the landing without incident. Wrote up the problem and called maintenance.

A ferry permit was issued, and the crew departed for the maintenance base. At the maintenance base the landing gear did not extend normally. The crew declared an emergency and pumped the gear down. There was a 3 down and locked indication when they landed. The landing, taxi in and shut down was [sic] uneventful.

Upon arrival, the captain was informed that the ferry permit was for a gear down ferry. I was the captain of this flight. After the landing I was surprised to learn about the restriction on the ferry permit. There are three factors I feel contributed to my oversight of the restriction. 1) After the initial landing, where the gear needed to be cycled once to work, I discussed the problem with the maintenance foreman. He actually asked me, "did you try to cycle the gear after it showed 3 down and locked?" No, was my response. The question left me thinking that he would have liked me to duplicate the problem. It led me to anticipate a maintenance test flight soon. This put me in the frame of mind that this was really a maintenance test flight and a ferry flight combined. 2) No one actually verbalized this was a gear down permit. For unknown reasons I didn't read the entire ferry permit. I don't know if I was interrupted or if I had so many other things on my mind at the time that I overlooked the last part of the gear down restriction. 3) I have never received any training on how to handle emergencies during ferry flights. I can only suggest some corrective actions so that someone else will not make such an oversight. 4) The permit should be referred to as a "gear down ferry permit." If flight control said they were sending me on a "gear down ferry," this would not have happened. If the permit had "gear down permit" at the beginning of it, I would have noted it. It actually started off reading "ferry for gear maintenance"! The part that read "ferry with gear down" was last. 5) Our airline does absolutely no training on the subject. Some ground and flight training should be conducted pertaining to single engine aircraft performance and capabilities during gear down operations.

Callback conversation with the reporter revealed the following information: The reporter stated that the gear problem occurred at Forbes airport prior to landing. There was no maintenance at Forbes, and the ferry permit was sent via teletype. The ferry flight was supposed to be gear-down from Forbes to MCI (Kansas City) where maintenance would have been done and a retraction test performed on the ground. The FAA required him to send in a report, and as of yet there have been no further repercussions. The reporter seems rather calloused about gear misbehavior. When asked if there were other gear problems he heard about, he said that about 2 weeks later, he had a gear indication abnormality. When the gear was lowered, the in-transit light came on, as it was supposed to, and then went out without the gear green lights coming on in a timely fashion. There was such a delay, the copilot and he had time to comment, "The gear-down lights did not come on." Then the green lights came on. But, it was the last flight of the day, and he did not write it up. The next morning he was deadheading on that aircraft's first flight, and the crew had to extend the gear by the alternate method. He did not know whether this incident involved the same aircraft of 2 weeks before.

Report synopsis
Flight crew raises gear on a gear-down ferry flight.

Aircraft involvement
Aircraft was involved in an anomaly (an unsafe or illegal event).

Fire Bottles Not Hooked Up—Missed by Captain during Checklist (ASRS Report #426822)

Date:	1999/01
Type of operation:	Air carrier, passenger flight
Aircraft:	High-wing, medium transport, between 30,001 and 60,000 lb
	Engines: Two turboprops
	Number of crew: Two
Flight crew:	ASRS reporter: first officer (F/O). 5500 hours, 250 hours within the past 90 days, 3000 hours in make/model. Air transport certificate.

Narrative (unedited, ASRS reporter's own words)
We had an originating flight out of a maintenance base. Captain did his flow (systems checks) while I did my preflight routine. He called for originating checklists which were completed. Later on, a different crew got

the airplane and noticed the fire bottle lights did not illuminate. On further inspection, the cannon plugs were not on the bottles, probably since it left maintenance base. When we did originating checklist and I called for fire detection system check, he said "complete." I would have to assume that he did do the check and did it correctly. Nowhere in our manual does it say a F/O (first officer) must get in the left seat and double check all the captain's system checks.

Report synopsis
Captain of a de Havilland Dash 8 (DA8) missed observing the engine fire extinguishing lights during the originating checklist, since it was later found that the fire bottles were not hooked up.

Aircraft involvement
Aircraft was involved in an anomaly (an unsafe or illegal event).

Summary and Assessment: Missed Checklist or Preflight Items

Missing checklist or preflight items can result in an unsafe flight, as illustrated by the six ASRS reports in this chapter.

    1. In ASRS report #126630, a corporate pilot did not check the fuel quantity following maintenance on the aircraft. In a subsequent test flight, the aircraft ran out of fuel, resulting in an off-airport landing.

    2. In ASRS report #117625, the flight engineer did not discover the downlock pin in landing gear during preflight. Maintenance failed to remove the pin. Aircraft

took off and could not retract the gear. Contributing factor was rush for on-time departure.

3. Damage to tail of aircraft was not noticed during preflight in ASRS report #144479 due to darkness. Damage was discovered by maintenance in hangar with hardstands and in daylight.

4. Although not necessarily a checklist item, a pilot-in-command retracted the landing gear on a ferry flight with a gear-down restriction in ASRS report #297664. The PIC did not notice the restriction on the ferry permit.

5. Pilot-in-command of a corporate aircraft used an abbreviated checklist on a test flight after maintenance instead of the full checklist. Thus, a malfunctioning autopilot was not detected until in flight (ASRS report #133799).

6. Captain missed observing the engine fire extinguishing lights checklist item. Fire bottles were not hooked up.

Although these are six different variations of checklist and preflight problems, they all come down to the same result:

Note: *Checklist and preflight functions must be adhered to.*

Checklist and preflight functions are generally covered adequately in training programs. Granted, there are always extenuating circumstances involved in missing checklist items such as pressure for on-time departure, distraction because of other tasks, excessive familiarity with the checklist, thus verbalizing a check without actually completing the checklist procedure, and stress or fatigue.

The human factor involved here is self-discipline.

There are all sorts of reasons why lack of self-discipline can be justified, and there are many organizational situations where supervisory imposition of discipline is not easy or effective. There is only one solution to either case. *That solution is integrity*—a quality within that maintains the state of order and submission to rules and authority without the need for anyone besides the pilot and him- or herself. In other words, follow established procedures—always.

See Chap. 10 for further discussion and assessment.

6

Flight Crew Misunderstanding of Aircraft Systems Resulting in Illegal or Unsafe Flight

The turbine-powered airplane is an assembly of fuel, hydraulic, electrical, and mechanical systems installed in the airplane's static components, such as the wings and fuselage. It is these active systems that provide maintenance and the pilot with the most problems in the operation of the airplane.

The Appendix provides an overview of a typical turbojet powered airline airplane, including its systems.

Jet Engine Overtemperature during Start (ASRS Report #110722)

Date: 1989/05
Type of operation: Air carrier, passenger flight
Aircraft: Low-wing, medium-large transport, between 60,001 and 150,000 lb

Engines: Two turbojets

Number of crew: Two

116 Chapter 6

Flight crew: ASRS reporter: captain, pilot-in-command. 13,400 hours, 150 hours within the past 90 days, 2300 hours in make/model. Air transport certificate.

Narrative (unedited, ASRS reporter's own words)
While starting #2 engine, EGT (exhaust gas temperature) went above maximum temperature. **Contributing factor:** We are operating 12 aircraft and have 10 different cockpit configurations. In one day we may start with one then go to a different model and then to another model. Several of our aircraft have different model engines on the same aircraft. **Example:** One of our aircraft has a brand X engine with a starting temperature of 420 and the other engine with a starting temperature of 550.

Report synopsis
During start, jet engine overtemperatured (beyond limits for that engine). Engine was shut down and maintenance notified.

Aircraft involvement
Aircraft was involved in an anomaly (an unsafe or illegal event).

Transport Flew with Only One of Two Ignition Systems Working When Both Are Required (ASRS Report #152571)

Date: 1990/07
Type of operation: Air carrier, passenger flight

Flight Crew Misunderstandings

Aircraft:	Low-wing, medium-large transport, between 60,001 and 150,000 lb
	Engines: Two turbojets
	Number of crew: Two
Flight crew:	ASRS reporter: captain, pilot-in-command. 9000 hours, 200 hours within the past 90 days, 500 hours in make/model. Air transport certificate, flight engineer rating.

Narrative (unedited, ASRS reporter's own words)

Due to a misunderstanding about fueling the aircraft (which has one fuel gauge inoperative), we pushed back from the gate about one hour late. When we tried to start #1 engine on left ignition got no ignition. Followed no start procedure, then started #2 engine on right ignition. Went back and started #1 on right ignition—all OK. Since we had one good ignition system which is legitimate for the other jets I've flown we were both satisfied. Didn't bother to call maintenance on the radio. Just flew to the destination and radioed in the problem en route. Maintenance supervisor met us at the gate and politely advised that on this aircraft, both ignition systems are required to be working! **Moral:** Take the time to do things by the book, no matter how late you are.

Report synopsis

Flight was late due to fuel problem. Flight crew found only one ignition system working. The flight continued. After review with maintenance, it was determined that both ignition systems are required on this aircraft.

Aircraft involvement
Aircraft was involved in an anomaly (an unsafe or illegal event).

Summary and Assessment: Flight Crew Misunderstanding of Aircraft Systems Resulting in Illegal and/or Unsafe Flight

Of the 79 ASRS reports in this book, only two involve misunderstandings of aircraft systems. Although these two ASRS reports have been singled out, illegal and/or unsafe flight due to misunderstanding of the aircraft systems is an underlying factor in most ASRS reports. This is discussed in greater detail in Chap. 10.

In these examples, the problem resulted from the pilot's operating different aircraft configurations without referring to the aircraft operations manual for familiarization with the individual aircraft systems.

In ASRS report #152571, the pilot-in-command started the engine on #2 ignition system with the #1 ignition system inoperative, which was legal on previous aircraft he had flown. However, on this aircraft, both ignition systems were required to be operational, resulting in an illegal and possibly unsafe flight. Old habits are difficult to break.

PIC of medium-large transport experienced an EGT overtemperature. This air carrier has different cockpit configurations as well as different engines with various EGT limits on the same aircraft (ASRS report #110722).

Flight Crew Misunderstandings

What solutions and/or lessons can be learned? Since the pilots cannot change the airline's mix of aircraft, pilots must take the time to review the individual aircraft operating manual and checklists before starting engines and preparing for flight. This is essential to make sure, among other things, that the pilot knows the engine starting procedures and operating limits.

7

Flight Crew Not Familiar with or Not Using Proper Abnormal or Emergency Procedures

In accordance with FAR part 91.503, for large and turbine-powered multiengine airplanes, each emergency checklist procedure must cover the following, as appropriate:

1. Emergency operation of fuel, hydraulic, electrical, and mechanical systems
2. Emergency operation of instruments and controls
3. Engine-inoperative procedures
4. Any other procedures necessary for safety

Based on this FAR, the aircraft manufacturer, together with the airline operating its airplanes, has developed detailed emergency and abnormal procedures. All pilots are exposed to these procedures during training sessions.

Fuel Imbalance Condition Misdiagnosed during Flight as Fuel Leak (ASRS Report #107424)

Date: 1989/03

Type of operation: Air carrier, passenger flight

Aircraft: Low-wing, wide-body transport, over 300,000 lb

Engines: Two turbojets

Number of crew: Two

Flight crew: ASRS reporter: captain, pilot-in-command. 16,100 hours, 205 hours within the past 90 days, 600 hours in make/model. Air transport certificate.

Narrative (unedited, ASRS reporter's own words)

Arrive on flight deck at XA PST for XB departure. Maintenance technician was in left seat preparing to run #2 engine to check for fuel leak, because high pressure fuel line had been changed on #2. Technician started engine and was told to immediately shut down because of large fuel leak.

Was told to expect 3 hour delay, but 30 minutes later was told problem was fixed, and we departed approximately one hour past schedule. Climb and initial cruise were normal, and I was especially observant of fuel quantities to make sure we did not have a recurring fuel leak. We climbed on center fuel tank, and it ran out approximately one hour 15 minutes into flight. At approximately one hour 30 minutes into flight, a fuel configuration warning light appeared on overhead panel, along with associated emergency message on instrument panel. I checked fuel gauges and found #2 main tank approximately 2000 lbs. lower than right main. I immediately assumed we had redeveloped a leak in right engine. I told F/O (first officer) to balance fuel and sent a phone patch through ATL (Atlanta) radio to our dispatch office. Meanwhile, I called

Abnormal or Emergency Procedures 125

first CSR (cabin service representative) from cabin and told him to look behind right engine/wing for any evidence of fuel leakage/contrail, etc.

After that I called center and advised them that we might need to divert. First CSR came back on flight deck and said he could see nothing from window, but I saw that fuel had decreased by another 300–400 lbs. in right tank. I decided we did probably have fuel leak and decided to secure right engine to decrease threat of fire. I requested and received clearance direct to Wichita, KS and began descent. At this time we completed engine shutdown checklist. An uneventful single engine landing followed on runway 19R at Wichita.

Once on the ground, no evidence of a fuel leak was found, and after I compared block in fuel with fuel burn, I concluded that we had just had a fuel asymmetry problem, and the engine had been shut down unnecessarily. As to what caused the asymmetry, I assume that the fuel crossfeed had been missed on F/O's prestart flow, and I had not caught it, or crossfeed had been inadvertently selected in flight or possibly a problem with fuel crossfeed valve itself or in fuel quantity indication. Whatever the cause, I believe my misdiagnosis of problem was certainly colored by the earlier fuel leak on ground.

Report synopsis
Because of a previous maintenance problem with leaking fuel, flight crew misdiagnosed a fuel imbalance condition as a fuel leak, shut down the suspected engine, and diverted to a nearby airport.

Aircraft involvement
Aircraft was involved in an anomaly (an unsafe or illegal event).

Transport Lands with One Gear Up Due to Disabled Uplock Pin Actuator (ASRS Report #154709)

Date: 1990/08
Type of operation: Air carrier, passenger flight
Aircraft: Low-wing, medium-large transport, between 60,001 and 150,000 lb

Engines: Two turbojets

Number of crew: Two

Flight crew: ASRS reporter: captain, pilot-in-command. 8000 hours, 220 hours within the past 90 days, 1000 hours in make/model. Air transport certificate, flight engineer rating.

ASRS Reporter #2: First officer (F/O), 3500 hours, 210 hours within past 90 days, 700 hours in make/model. Air transport certificate.

Narrative (unedited, ASRS reporter's own words)

According to the logbook the aircraft had not been flown for 10 days. Maintenance, including a new interior, had been done at OAK (Oakland, CA). I did not discover this until reviewing the log book. Since it was F/O's (First Officer's) leg, I did the exterior while he started the cockpit setup.

The trip was uneventful until the approach phase. During the approach at BUR (Burbank, CA) with the

Abnormal or Emergency Procedures

F/O at the controls, the right main gear "unsafe" red light came on and stayed on during gear extension with no green down and locked light. We were about 8–10 miles out at about 3000–4000 feet and 180 kts IAS (indicated air speed) with 5 degrees flaps. I recycled gear with the same result. After F/O made a go-around, I made a PA announcement regarding the gear light problem while being vectored in the BUR area.

We cycled the gear several times. After talking to maintenance, we cycled the gear at 220 kts. (at 10000 ft.), the maximum gear extension speed. Initially, I suspected a gear light problem, since I had seen similar problems while on the light transport which were corrected before landing. The F/O went to the cabin to check the gear down locks through the view port, and the first F/A (flight attendant) was called to the cockpit to discuss an emergency evacuation preparation. After the F/O came back, he called maintenance, then reentered the cabin. He confirmed the right main gear was up. He resumed his duties and prepared for a landing at LAX (Los Angeles).

The F/As were very professional and did not "chime" us or in any way disrupt our procedures. They were very patient and probably understanding of the workload in the cockpit. When asked to prepare the cabin, they did so in minimum time and reconfirmed the brace signal. When finished, they informed me that they were returning to their seats for the landing.

An air traffic controller told us that another aircraft had seen the right gear up with the others down. Also, later while passing near BUR, the tower saw the right gear up. At no time did I consider the landing to be life threatening even with one main gear up because of the engine size. When the right main gear was confirmed up, we elected to go to LAX instead of a planned pass by BUR tower. The F/O and I set up for LAX and he began checklists including "partial gear or gear up landing checklist." LAX was chosen for its long runways and emergency equipment. I asked that emergency equipment be standing by and assumed that they knew from BUR approach control that our right gear was up. On our way to LAX, I informed the passengers of the problem and told them that the landing should not be a major problem because of the large diameter engine, but we would evacuate the aircraft when it came to a stop.

When the right engine did contact the pavement, directional control was maintained and deceleration was fine until at a very slow speed, the aircraft began turning to the right with some shaking. I shut down both engines to stop the aircraft, set the brake, pulled the fire handles (including APU) and rotated the #2 fire handle to discharge the extinguisher. I suspected some friction fire where the engine was in contact with the runway. Evacuation went smoothly. In fact by the time (only a few minutes) F/O and I entered the cabin, the F/As had all the slides inflated and all but about 10 passengers off. These were waiting to go out the left wing exit so we redirected them to the front doors. Then the crew exited.

Callback conversation with reporter revealed the following information: Callback completed to reporter #2 (first officer). Mechanics discovered that when the aircraft was serviced prior to flight, the right gear strut was serviced and a mechanic had left a flashlight resting above the gear uplock actuator. When the gear recycled, the flashlight was crushed by the actuator, and the actuator was dented. This prevented actuator from being able to pull the uplock pin.

After the incident, the flight crew reenacted the scenario in the simulator. The company suggested that the crew might have taken more time to confer with the company in order to find a solution.

During callback, the F/O said that both he and the captain had recently come from flying large aircraft, and that their perception of just what constituted a critically low fuel state was distorted by their previous experience in other aircraft. The flight landed with 6000 lb of fuel remaining, and the company thought that, had the flight crew used an additional 2000 lb, they would have gained perhaps 15–20 minutes of time for troubleshooting and preparation. As it was, the F/O said that once the decision was make to go to LAX, the fight crew was so busy in trying to prepare for the landing that they completely forgot about the gear problem. The flight crew was unable to complete the preliminary checklists prior to landing, as they were entirely consumed with trying to configure the aircraft, comply with ATC instructions, pull out their approach plates, watch out for traffic, and reprogram the FMC (flight management system).

Only about 5 minutes elapsed between the decision to land at LAX and touchdown. Had the crew been able to confer with the company more extensively, the company had a special team that could have been hooked up across the country with direct communication between them and the flight crew for the purpose of troubleshooting and preparing for an emergency landing. That team would have suggested that the flight crew attempt to deploy the landing gear by using the T-handle located in the cockpit floor.

Although the flight crews are not told this during their training, the T-handle actually had a direct connection to the gear unlock pin, and the possibility exists that the crew could have manually extracted it. There is no reference to using the T-handle in the gear-up landing procedure, because its use is usually predicated on a loss of hydraulic power, which did not exist in this circumstance.

130 Chapter 7

The company had only recently promoted the use of the troubleshooting team in its flight crew newsletter, but the reporter had not yet read it. The F/O suggested that the company more clearly mark the location of the landing gear viewport in the cabin, as the F/O was unable to locate it in his first attempt. F/O believes that the company will make changes in this regard.

Report synopsis
Flight crew of medium-large transport aircraft arriving at BUR was unable to lower right main landing gear due to flashlight left near gear uplock pin actuator by maintenance, thus disabling the actuator. Flight diverted to LAX where it landed with right main gear up.

Aircraft involvement
Aircraft was involved in an anomaly (an unsafe or illegal event).

Uncommanded Rudder Displacement, Crew Maintains Level Flight Using Standby Rudder on "B" Hydraulic System (ASRS Report #429547)

Date:	1999/02
Type of operation:	Air carrier, passenger flight
Aircraft:	Boeing B737-200, low-wing, medium-large transport, between 60,001 and 150,000 lb
	Engines: Two turbojets
	Number of crew: Two

Abnormal or Emergency Procedures 131

Flight crew: ASRS reporter: first officer (F/O), pilot-in-command (PIC). 15,000 hours, 200 hours within the past 90 days, 2000 hours in make/model. Air transport certificate.

Narrative (unedited, ASRS reporter's own words)

Flight from MCO (Orlando, FL) to BDL (Hartford, CT/Springfield, MA), 33,000 feet, 0.67+ mach normal cruise. Experienced uncommanded roll to right. First noted by displacement of control wheel to left with no pilot input. Took control of aircraft. Noted that the rudder pedals displaced to right, right rudder forward and unable to center. Maintained close to level flight with considerable left aileron input. Declared emergency, complied with procedures in QRH (quick reference handbook). Both pilots could not correct rudder. When "B" system hydraulic switch placed to standby rudder, rudders centered and controllable. During approach, noted vibration in rudder pedals and some rudder 'kicks.' Uneventful landing BWI (Baltimore, MD), runway 33L. Made appropriate logbook entry.

Supplemental information: I declared an emergency with ATC. Immediate action items consisting of autopilot, autothrottles, and yaw damper were turned off. Both pilots attempted to move jammed rudder pedals. No desired movement observed. QRH procedures complied with. Rudders were freed up when "B" system hydraulic system control switch was placed to standby rudder position.

Callback conversation with reporter revealed the following information: The reporter said that the captain was flying the aircraft on autopilot at the time of the rudder displacement, and the reporter attempted to assist when the right rudder pedal went full travel forward. Then as the flight crew started through the uncommanded rudder input checklist, this reporter assumed control of the aircraft. The reporter has spoken to maintenance personnel, and they have been unable to discover the cause for the rudder displacement. The aircraft had one of the new rudder control valves installed, and it was in proper working order according to maintenance.

Additional callback conversation with the reporter revealed the following information: This reporter said that with both flight crew members pushing on the opposite rudder pedal, they were unable to correct the right rudder displacement. However, he was able to maintain level flight by using almost full aileron input. He said that as soon as he placed the B system hydraulic switch to the standby rudder position, the rudder became controllable again.

Report synopsis
This B737-200 aircraft flight crew experienced an uncommanded rudder displacement that caused the aircraft to roll to the right. Initially, the flight crew were unable to overpower the uncommanded input but were able to maintain approximately level flight by using the B hydraulic system standby rudder.

Summary and Assessment: Flight Crew Not Familiar with or Not Using Proper Abnormal or Emergency Procedures

A B737 flight crew (ASRS report #429547) experienced an uncommanded rudder displacement that caused the aircraft to roll to the right. The flight crew did not use abnormal or emergency procedures for this potentially disastrous condition, because such procedures did not exist. However, by operating as a team and using their skills, knowledge, and expertise, they managed to regain control of the aircraft and diverted to a nearby airport for a safe landing. See *Aviation Week & Space Technology,* March 1, 1999, page 41, for an article on this problem.

In ASRS report #107424, the flight crew misdiagnosed a fuel asymmetry as a fuel leak because of a previous maintenance problem with leaking fuel and shut down an engine unnecessarily. The fuel crossfeed on the prestart checklist had been missed.

Flight crew made an emergency landing and damaged the aircraft with one landing gear not down and locked. Flight crew was not aware of a procedure to manually remove the gear unlock pin (ASRS report #154209).

These solutions were found and lessons learned:

1. Crew coordination, as well as the expertise and knowledge of aircraft systems, allowed the flight crew of a B737 to maintain control of the aircraft after an uncommanded rudder displacement when no abnormal or emergency procedures existed for this event.

2. In ASRS reports #154209 and #107424, the flight crews did not use the complete checklist. However, in both of these incidents, the underlying cause was a lack of sufficient understanding of the airplane's systems.

3. In ASRS report # 154209, in particular, the reporter indicated that he had no specific training regarding the procedure to manually remove the gear unlock pin since it was included in the hydraulic system emergency

procedure, not the "partial gear, or gear-up landing checklist."

These ASRS reports again emphasize that the best preparation for system emergencies is to be thoroughly familiar with the airplane's systems. This should include self-study beyond the airline's training sessions.

See Chap. 10 for further discussion of this subject.

8

Flight Crew Not Checking with Maintenance When Discrepancy Exists

It is a violation of the FAR, as well as company procedure, for a flight crew to accept an aircraft for flight with open maintenance items. A summary and assessment of the ASRS reports included in this chapter is presented in the last pages of the chapter.

Subsequent Departure without Maintenance Sign-Off after In-Flight Engine Shutdown and Adding Oil to Gearbox That Overheated (ASRS Report #90741)

Date: 1988/7
Type of operation: Air carrier, freight
Aircraft: Low-wing, medium transport, between 30,001 and 60,000 lb

138 Chapter 8

Flight crew:

Engines: Two turboprops
Number of crew: Two
ASRS reporter: captain, pilot-in-command. 6300 hours, 135 hours within the past 90 days, 2000 hours in make/model. Air transport certificate.

Narrative (unedited, ASRS reporter's own words)

During cruise, I got an accessory gearbox light (red). Procedure calls for engine shutdown. We secured the engine and notified ATC (air traffic control). Then proceeded to our destination LCK (Lunken, Cincinnati, OH) which was 100 NM (nautical miles) away. We made an uneventful landing. I secured the plane and called my flight following department at DFW (Dallas/Ft. Worth) who then had maintenance call me. Maintenance suggested I check the gearbox oil; add if needed, then run up the engine. I added 1 quart of oil and ran up the engine.

The light didn't come back on. The problem appeared to be solved. We have no local maintenance in LCK so upon DFW telling me the problem was now solved, we subsequently departed and returned to Richmond, our base. Upon arrival I entered the shutdown in the air carrier book and advised our local maintenance at RIC (Richmond/Williamsburg, VA) and DFW. The problem was corrected and signed off. **My problem:** I should have stayed on the ground in LCK after the shutdown and forced our DFW maintenance to arrange for an inspection and sign off prior to departure. I have since met with our director of operations, chief pilot and maintenance to confirm this and told them that I will stay on the ground next time.

Report synopsis
Air carrier, medium transport, had in-flight engine shutdown due to high accessory gearbox temperature. Flight crew subsequently departed after adding oil to gearbox but without maintenance sign-off.

Aircraft involvement
Aircraft was involved in an anomaly (an unsafe or illegal event).

Aircraft with Skin Damage Flown without Proper Inspection (ASRS Report #92184)

Date:	1988/08
Type of operation:	Air carrier, passenger flight
Aircraft:	Low-wing, medium-large transport, between 60,001 and 150,000 lb
	Engines: Two turbojets
	Number of crew: Two
Flight crew:	ASRS reporter: captain, pilot-in-command. 6000 hours, 200 hours within the past 90 days, 2000 hours in make/model. Air transport certificate.

Narrative (unedited, ASRS reporter's own words)

At the first landing of a 2 leg trip, the aircraft was hit by a baggage cart as it was being pulled up to the aircraft. It was hit several times as the cart bounced off the stringers. There was no mechanic on duty so ramp called one at home. He said the aircraft would be OK if the skin was not punctured. Ramp then looked at aircraft and then notified me, at this point, that the skin had not been punctured. I decided to look myself. The light was poor, and it was pouring rain at the time.

It appeared to me that the skin was not punctured, but it was dented badly—but no more than I had seen before. Since there appeared to be no punctures, and there was no mechanic to sign the maintenance log off, I decided not to write it up until returning to base—we were already late. The return trip was normal. On arrival, I requested maintenance to inspect aircraft. They did find one small crack, which in sunlight and dry skin was visible. The aircraft was removed from service for repair. In retrospect, the aircraft should have been written up and left on the ground as the safest course of action.

Callback conversation with reporter revealed the following: Reporter said that this incident took place at Hilo and that he has heard nothing further about it from anyone.

Report synopsis

Aircraft with skin damage is flown without proper maintenance inspection.

Aircraft involvement

Aircraft was involved in an anomaly (an unsafe or illegal event).

Electrical Problems Activate Warning Circuits during Aborted Takeoff and after Second Takeoff with No Maintenance Activity (ASRS Report #101677)

Date: 1989/01
Type of operation: Air carrier, passenger flight
Aircraft: Low-wing, medium-large transport, between 60,001 and 150,000 lb

Engines: Two turbojets

Number of crew: Two

Flight crew: ASRS reporter: first officer (F/O). 12,500 hours, 150 hours within the past 90 days, 1000 hours in make/model. Air transport certificate, flight engineer rating, certified flight instructor.

Narrative (unedited, ASRS reporter's own words)

On the takeoff roll at SRQ (Sarasota/Bradenton, FL) we heard a loud noise coming from the main CB (circuit breaker) panel. We also got the master caution light, forward cabin door light and the master fire warning light with no other fire indications. We aborted at 90 kts. and cleared the runway. We started to troubleshoot. We had not lost a generator but found 5 CBs popped on different systems.

Since we smelled no smoke, we thought it was a freak electrical surge. We reset the breakers per company policy and tried to contact maintenance by commercial radio but were unsuccessful. We do not have maintenance at SRQ, and the breakers did not pop again. We elected to depart SRQ. The second takeoff was normal, and we experienced no further problems for the next hour of flight.

Then essentially the same thing happened again. Fearing an electrical fire and because it happened twice, we elected to declare an emergency and land at the nearest suitable airport CHS (Charleston, SC). We called our maintenance center at TUL (Tulsa, OK) enroute [sic] to explain what happened on the ground at SRQ and to ask them if they had any further suggestion. They could offer no solution but would notify RDU (Raleigh Durham, NC) maintenance to stand by to check the plane.

We grounded the plane upon arrival at CHS. At CHS we noticed that at least 10 CBs had popped. Emergency equipment was standing by. We found out the next day from maintenance that a power transfer relay had fused.

Human performance: Company's crew resource management is really given emphasis and F/Os (first officers) are encouraged to participate in decision making. I felt like we had taken every precaution possible and felt free to offer suggestions.

Back 7 days later with 20/20 hindsight, perhaps we should have returned to the gate in SRQ, but we did not have maintenance there, and I didn't feel that a mechanic from local FBO (fixed base operator) unfamiliar with the medium large transport could offer much input. Perhaps our desire to deliver a full load of people clouded our judgment.

Supplemental information: After attempts to call Tulsa tech with no success, reset CBs, started APU (auxiliary power unit), ran up engines to 1.6 EPRs (engine pressure ratio) and had no adverse indications. Elected to attempt another takeoff with APU running. Decision to attempt second takeoff at SRQ was made because of reputation of aircraft and glitches that seem to remedy themselves and lack of maintenance at SRQ, full load of passengers and company pride to get them to their destination.

Report synopsis
Air carrier, medium-large transport, aborted takeoff because of electrical problems that activated several warning circuits. Problem recurred at cruise altitude after second takeoff was made without any maintenance activity. Emergency was declared, and flight was diverted to nearest suitable airport.

Aircraft involvement
Aircraft was involved in an anomaly (an unsafe or illegal event).

Pilot Performed Unauthorized Minor Maintenance Suggested by FAA Inspector (ASRS Report #108519)

Date:	1989/04
Type of operation:	Air taxi, preflight
Aircraft:	Low-wing, small aircraft, less than 5000 lb, retractable gear
	Engines: One reciprocating
	Number of crew: One

Flight crew: ASRS reporter: single pilot. 2350 hours, 200 hours in the past 90 days, 300 hours in make/model. Air transport certificate. General aviation pilot flying for compensation at time of occurrence.

Narrative (unedited, ASRS reporter's own words)

I was ramp checked by an FAA inspector. He noticed that the navigation light holder on the right wingtip was slightly loose. He pointed it out to me and said, "If you have a screwdriver, you should tighten that up. I'm afraid you may lose it in flight." I had a screwdriver in my pocket and did as he suggested and thanked him. He did not mention any other discrepancies and said goodbye.

Upon returning to our office, I was advised by our director of operations and chief pilot that I was, in fact, performing maintenance and would receive a notice of violation for the action. I tightened the screw believing I was complying with the FAA inspector's wishes. Knowing that regulations do not allow a pilot under part 135 to perform even such trivial maintenance, I now realize I should have refused to tighten the screw.

Report synopsis

Pilot on preflight check with air carrier inspector performed minor maintenance at suggestion of air carrier inspector.

Aircraft involvement

Aircraft was involved in an anomaly (an unsafe or illegal event).

Aircraft Departed with Open Maintenance Item in Logbook (ASRS Report #114248)

Date: 1989/06
Type of operation: Commuter air carrier, passenger flight
Aircraft: Low-wing, light transport, between 14,501 and 30,000 lb
Engines: Two turboprops
Number of crew: Two
Flight crew: ASRS reporter: first officer (F/O). 1800 hours, 200 hours within the past 90 days, 750 hours in make/model. Commercial license with instrument rating, flight engineer rating.

Narrative (unedited, ASRS reporter's own words)

We had an XB departure, and our aircraft didn't block in until XA after rushing to get out on time. One of the management pilots, jumpseating [sic] on company business, pointed out a write-up in the aircraft logbook which had not been signed off. The captain claims to have been told by the crew getting off that there were no write-ups. I just plain didn't look at it. Fortunately, the write-up concerned the APU (auxiliary power unit) which we MEL'd (minimum equipment list) by phone at the destination. The problem arose due to a quick turn, aircraft swap, and a failure to thoroughly complete a checklist item. "Logbook and manuals...checked/on board."

After the walk around, getting ATIS (automatic terminal information service) and a clearance, I had checked the log for a current preflight/daily maintenance inspection, checked the previous write-ups for a history, checked the deferred page, and entered the crew names on the first sheet of the day but failed to lift the first and look at the second sheet for "today's" write-ups history (or open). Factors contributing were, first, the captain's failure to check the log. Second, the previous crew either did not call maintenance or if they had, maintenance never had time to respond. Third, although not a written policy, in practice, when a crew is getting off an aircraft, leaving an open write-up, tell the next crew or leave the log in a conspicuous place open to the second page with the gripe showing for the next crew or maintenance to see. In this case, the log was stowed in a slot between the cockpit seats. Fourth, the captain is not one of the favorites to fly with, for various reasons, which has a negative effect on quality performance.

To prevent this from recurring, I have reviewed the expanded version of the receiving checklist, made my own checklist of duties for getting a new aircraft, and will absolutely refuse to be rushed.

Report synopsis
Aircraft departed with open maintenance item in logbook.

Aircraft involvement
Aircraft was involved in an anomaly (an unsafe or illegal event).

Departure without Clearance Regarding Maintenance Problem (ASRS Report #134560)

Date: 1990/01
Type of operation: Air carrier, passenger flight

Not Checking with Maintenance

Aircraft:	Low-wing, medium-large transport, between 60,001 and 150,000 lb
	Engines: Two turbojets
	Number of crew: Two
Flight crew:	Captain, pilot-in-command. 6569 hours, 193 hours within the past 90 days, 227 hours in make/model. Air transport certificate, flight engineer rating.

Narrative (unedited, ASRS reporter's own words)

During engine start, #2 engine would not start when using the right ignition. It started OK using the left igniter. A message was sent to maintenance via radio. Due to concern about the hold overtime of the deicing fluid, which had been sprayed on the aircraft about 15 minutes earlier, we departed without proper response from maintenance.

I feel that my concern about the deicing fluid caused me to forget that I was waiting for a message via radio from maintenance. This did not affect the flight but was a violation of our standard operating procedure.

Report synopsis

Air carrier crew departed without receiving clearance regarding maintenance problem from company personnel.

Aircraft involvement

Aircraft was involved in an anomaly (unsafe or illegal event).

Captain Got Maintenance to Sign Off on Engine Check That Had Not Yet Been Done (ASRS Report #138343)

Date: 1990/02
Type of operation: Air carrier, passenger flight
Aircraft: Low-wing, wide-body transport over 300,000 lb
Engines: Three turbojets
Number of crew: Three
Flight crew: First ASRS reporter: first officer (F/O). 13,000 hours, 250 hours within the past 90 days, 1200 hours in make/model. Flight engineer rating, commercial license with instrument rating.

Second ASRS reporter: captain, pilot-in-command. 15,000 hours, 225 hours within the past 90 days. Air transport certificate, flight engineer rating, certified flight instructor.

Narrative (unedited, ASRS reporter's own words)

Inbound flight arrival LGA (LaGuardia, NY) with a maintenance write-up. The #2 engine would not develop proper takeoff and climb power. The outbound flight was given advise time while maintenance analyzed the #2 engine power problem. At the same time, the captain was told by two LGA maintenance supervisors that the

#2 engine's fuel flow regulator was being adjusted for maximum takeoff power. The engine would then be run up and checked for the flight.

Captain then suggested to the LGA maintenance that they need not take the time or delay for a run up check of the engine and urged the passengers be boarded ASAP. He told maintenance supervisors that he would do the necessary #2 engine power check on takeoff roll, and if the power indication was not adequate, he would abort takeoff and return to the gate. The passengers boarded flight parked at gate. LGA maintenance sign off on the #2 engine write-up (i.e., inbound flight same aircraft) but stating the adjustment of the #2 engine's fuel flow regulator, engine run up and power checked OK.

At this point, F/O (first officer) (I) advised the captain that a takeoff attempt without prior engine run up and check by maintenance would be an unsafe maneuver and poor judgment on the part of the crew. Captain then told me he was willing to take a 50/50 chance so as not to incur further delay. I then expressed my concern to captain about any false statements or statement in the aircraft logbook by LGA maintenance. That is, the statement referring to #2 engine being run up and checked for proper takeoff power indications by maintenance. I advised the captain that in making such statement of fact and not following specified procedures was a felonious action and one that could put us all at risk for prosecution.

He then told me not to worry; he had talked with LGA maintenance supervisor and was assured that if the FAA showed up they could indeed say that the #2 engine was run up and checked prior to takeoff as per signoff in the logbook. After my several attempts to dissuade captain from acting on an improper procedure, I acquiesced to his decision and flight departed gate with crew and passengers. Flight was cleared for takeoff on runway 31 LGA. On initial takeoff roll all three power levers were advanced for maximum takeoff power. The #2 engine failed to reach the required takeoff power and the decision was to abort takeoff.

LGA tower was advised of aborted takeoff and informed of exact cause, that being inadequate power of the #2 engine. No emergency assistance was requested or determined to be needed. We proceeded back to the gate where it was [sic] met by LGA maintenance and appropriate logbook entry was made. #2 engine's maximum takeoff indications were approximately EPR 1.410, N1, 85%. Passengers of flight were deplaned, and the flight was cancelled as a revenue flight.

Above is first officer's (F/O) report. Below is the captain's report:

It has been alleged that I accepted aircraft knowing there was an improper logbook entry. I emphatically deny the allegation. The item in question concerned the #2 engine's ability to provide full takeoff thrust on the previous takeoff. The corrective action by maintenance was to adjust the fuel control. It has been alleged that maintenance should have confirmed the fix by a full power run up of the engine. This was not important to me, since I was assured by maintenance personnel that the engine would now perform up to specifications. Furthermore, I would make the decision (as I do on every takeoff) to accept or not to accept the engine's performance.

My decision would be based on the engine's performance on that takeoff regardless of maintenance's evaluation. The engine, in fact, did not perform as required, and the takeoff was aborted at about 65 kts. This was hardly a panic stop. In fact, I had to add power to get to the turnoff. My major concern with this event was, and is, my second-in-command's constant second guessing my decisions. While I am always open to other crew member's suggestions, there can only be one PIC (pilot-

Not Checking with Maintenance 151

in-command). The current emphasis on CRM (crew resource management) has been misinterpreted by some to mean that everyone has equal authority in the cockpit. This entire event would not even be worthy of mention if the pilot second-in-command had performed his duties as a proper copilot.

Report synopsis
Captain of outbound flight, a wide-body aircraft, got maintenance to "cook the books" on an engine maintenance snag encountered in inbound flight.

Aircraft involvement
Aircraft was involved in an anomaly (an unsafe or illegal event).

After Circuit Breaker Opened on a Previous Flight, Crew Failed to Make Logbook Entry and Get a Maintenance Write-Off before Next Departure (ASRS Report #149395)

Date:	1990/06
Type of operation:	Commuter air carrier, passenger flight
Aircraft:	Low-wing, light transport, between 14,501 and 30,000 lb
	Engines: Two turboprops
	Number of crew: Two
Flight crew:	ASRS reporter: captain, pilot-in-command. Air transport certificate.

Narrative (unedited, ASRS reporter's own words)

Upon arrival into the Houston area, we started our approach. When we extended the gear there were no indications of the 3 gear lights. We noticed that the gear indicator light CB (circuit breaker) was popped out. The copilot reset the CB, and the 3 gear lights were illuminated. We flew by the tower to ensure the gear was down, and they indicated that it was down. As a precautionary measure, I told the tower to have the fire trucks standing by.

We circled the airport and landed with no incident. On the ground, I called maintenance and was on the phone for approximately 5 minutes. I waited until approximately 5 minutes after departure time for a return phone call. I departed the aircraft to Austin and neglected to communicate with maintenance. Considering the passenger comfort and company schedule, we flew the aircraft to Austin. Since the aircraft seemed to be airworthy, being the only thing wrong was a CB that had popped out and was reset. The aircraft arrived in Austin, and there was nothing found wrong with it. Maintenance released the aircraft as airworthy.

Report synopsis

Commuter, light transport flight crew failed to make a logbook entry and get a maintenance write-off before departure. Gear warning circuit breaker had opened on the previous flight.

Aircraft involvement

Aircraft was involved in an anomaly (an unsafe or illegal event).

Crew Made Fuel Transfer When Fuel Leak Noticed without Notifying Maintenance (ASRS Report #158584)

Date: 1990/09
Type of operation: Air carrier, passenger flight
Aircraft: Low-wing, medium-large transport, between 60,001 and 150,000 lb

Engines: Two turbojets
Number of crew: Two

Flight crew: ASRS reporter: captain, pilot-in-command. 15,000 hours, 200 hours within the last 90 days, 4000 hours in make/model. Air transport certificate.

Narrative (unedited, ASRS reporter's own words)

We landed at Islip 10/90 having flown the same aircraft twice previously the same day with no fuel problems. This model medium large transport had a fuel capacity of 24649 lbs. We required 21000 lbs. for the flight to Atlanta. The fueler loaded the fuel 9000 lbs. left, 3000 center, 9000 lbs. right. It should have been loaded 8500 lbs. in each wing tank and 4000 lbs. in center. We noticed a small amount of fuel leakage near the L main gear. We traced the leak to a small access panel on top inner portion of the L wing. We transferred the fuel to the proper tanks, and the leakage stopped.

We took off about 10–15 minutes later with no evidence of any fuel leak. We had no fuel problems on the flight to Atlanta. At no time did we contact maintenance for assistance. In retrospect, I believe this was a mistake. We followed the "normal" procedure for ground transfer of fuel to the proper tank configuration, and this resulted in a "healthy" aircraft with no leaks. A complete follow-up would have been to have maintenance determine what caused the leakage in the first place. The fact that this was the last flight of a four-day trip may [have] influenced me to leave without determining the "root cause" even though I had assured myself that the aircraft was safe. Possibly the old "get homeitis" syndrome.

Report synopsis
Medium-large transport flight crew made a ground fuel transfer when a fuel leak was noticed from the left main tank. Maintenance was not notified, and transfer was made after departing the blocks.

Aircraft involvement
Aircraft was involved in an anomaly (an illegal or unsafe event).

Pilot Advised of Violation Because Some Maintenance Items Had Not Been Signed Off by a Mechanic (ASRS Report #168110)

Date: 1991/01

Type of operation: Instrument training flight in small general aviation aircraft

Aircraft:	Low-wing, fixed-gear, small aircraft, less than 5000 lb
	Engines: One reciprocating
	Number of crew: Instructor pilot and student
Flight crew:	ASRS reporter: general aviation instructor pilot. 4500 hours, 150 hours in the past 90 days, 3000 hours in make/model. Air transport certificate, certified flight instructor.

Narrative (unedited, ASRS reporter's own words)

My student and I went on an instrument training flight. We flew a small aircraft, which the flight school purchased one week ago. We had intended to remain VFR (visual flight rules) due to FSS (flight service station) information, but when we became airborne, we found the ceiling lower than reported, so we filed IFR (instrument flight rules) and completed the training flight by doing 2 ILS (instrument landing system) approaches at ALN (Alton, IL).

During the ILS approach, the G/S (glide slope) flag was showing and the G/S did not work, so we completed the approach as a LOC (localizer) approach. When we returned to the airport after the flight, 2 FAA inspectors met us. They informed me that I had flown an unairworthy airplane, because it had an open squawk on it when I took off. Apparently, the person who flew it before me had squawked an alternator light and intermittent #2 radio and had noted that the vacuum system should be checked. The FAA inspector, Mr. X, had told me earlier that I, as the pilot, could defer the squawks.

156 Chapter 8

Now he tells me that only a mechanic could defer the squawk on the alternator. I told him that I had checked all three squawks and found no problem with them. He said I should have had a mechanic check the alternator light even though it gave me no problem. The second FAA inspector, Mr. Y, said, "Ah ha! You got one!" to Mr. X. Mr. X had definitely told me earlier that I could defer a squawk.

Report synopsis
After the fact, a general-aviation small-aircraft instructor pilot is advised he had operated in violation of Federal Aviation Regulations, because some maintenance items had not been signed off by a mechanic.

Aircraft involvement
Aircraft was involved in an anomaly (an unsafe or illegal flight).

Aircraft Returns for Landing after Fire Warning Due to Fuel Leak after Installation of Fuel Nozzle (ASRS Report #173774)

Date:	1991/01
Type of operation:	Commuter air carrier, passenger flight
Aircraft:	Low-wing, medium transport, between 30,001 and 60,000 lb
	Engines: Two turboprops
	Number of crew: Two
ASRS reporter:	Ground crew

Narrative (unedited, ASRS reporter's own words)

The aircraft was in PVD (Providence, RI) for maintenance. At this time, a fuel nozzle change was done on the #1 engine. The aircraft dead headed [sic] to BDL (Hartford, CT/Springfield, MA), and then with passengers went to JFK (Kennedy, New York, NY) without incident. There was some confusion at this point, but I am led to believe that the captain did, while at JFK, report a suspected fuel leak from the #1 engine but was told by maintenance at that station that fuel leakage from the environment drain at the nacelle was normal after a nozzle change, and no further action was taken.

Upon takeoff, #1 engine fire alert activated. Engine was shut down and fire warning was confirmed. Fire bottle was discharged and fire was extinguished. Aircraft returned to JFK and landed without further incident. Damage to #1 engine and nacelle was extensive. Upon investigation, it was noted 3 fuel transfer tubes on one nozzle at 2 o'clock were improperly installed under the lock and clip and worked out of the fuel nozzle boss allowing fuel to spray into the nacelle. It is my belief that the majority of fault was human, i.e., the mechanic who performed the work, the inspector, and maintenance at JFK for not following up and checking the nacelle interior for fuel and to some degree, the captain for accepting their explanation and failing to notify maintenance control of his suspicion.

Also, to an even greater degree, engine manufacturers for poor design of the locking clip on the engine of which there are 4 modifications and the operator for failure to inform us that we were still using the older style that allowed improper assembly of fuel transfer tubes. To my knowledge, no corrective action had been taken yet.

Report synopsis
Mechanic reports that after maintenance, the aircraft had fire warning after takeoff and returned for landing. There was damage to the nacelle and engine because of a fuel leak after installation of a fuel nozzle.

Aircraft involvement
Aircraft was involved in an anomaly (an unsafe or illegal event).

Crew Proceeded with Takeoff without Consulting Maintenance after Aborting First Takeoff When They Noticed Reverser Pressure Light (ASRS Report #181587)

Date:	1991/06
Type of operation:	Air carrier, passenger flight
Aircraft:	Low-wing, wide-body transport, over 300,000 lb
	Engines: Three turbojets
	Number of crew: Three
Flight crew:	ASRS reporter: second officer, flight engineer (F/O) rating. 7200 hours, 230 hours within the past 90 days, 3000 hours in make/model.

Narrative (unedited, ASRS reporter's own words)

In position for takeoff, runway 8R. As we were cleared for takeoff, takeoff checklist final items completed. As power was brought up, all engine indications looked good (pilots' and flight engineer's instrument panels). At approximately 60 KIAS, I noticed the engine #2 reverser pressure light on the F/O (flight engineer) panel illuminated (pneumatic pressure to operate air driven reverser motor and remove reverser locking pins). We decided to discontinue the takeoff roll and try to find the cause of the problem. As power was reduced, reverser pressure light went out.

We taxied clear of the runway and attempted to duplicate the problem with a high power engine run up. We were unable. We decided not to consult our maintenance department, since that would involve taxiing back to the gate and taking an additional delay; we were already one hour late in departing. However, we did decide that if it happened again, we would, without question, taxi back and consult maintenance. It has come to my knowledge that the captain of this flight did not as yet file an incident report with the company, as is required by operation specifications. Depending upon the mood of the inspector, this could be a violation of the FARs as has happened to two other captains recently. So far, however, the FAA has not come a'calling—so far.

Report synopsis

An air carrier wide-body flight crew experienced a reverser pressure light and aborted takeoff, only to try again without consulting maintenance.

Aircraft involvement
Aircraft was involved in an anomaly (an unsafe or illegal flight).

Commuter Transport Returned to Airport Instead of Diverting to Airport (Recommended by Maintenance) after Getting False Warning on Master Panel (ASRS Report #182759)

Date:	1991/06
Type of operation:	Commuter air carrier, passenger flight
Aircraft:	High-wing commuter, medium transport, between 30,001 and 60,000 lb
	Engines: Two turboprops
	Number of crew: Two
Flight crew:	ASRS reporter: first officer (F/O). 3488 hours, 200 hours within the past 90 days, 1530 hours in make/model. Air transport certificate, certified flight instructor.

Narrative (unedited, ASRS reporter's own words)
Upon initiating a positive climb at lift off, landing gear was retracted. At that time, there was a master caution with an engine light illuminated on the central crew

alerting panel. Review of all engine conditions revealed nothing. We informed MSN (Madison, WI) departure that we had a maintenance indication and would like to hold VFR and check it out. Still unable to determine why we still had the light, we informed departure we would like to hold on a radial of Madison VOR while we talked to our company maintenance through relay with STL (St. Louis, MO).

Maintenance asked us to confirm "no abnormal engine conditions." We did. He said to bring it back to STL. We disagreed and said that we would not unless he could explain what was causing the caution. He said he couldn't—that it was probably an erroneous light. We elected to return to MSN and write up the aircraft. Upon landing, captain was instructed to call tower. The FAA was on the phone to talk to the captain.

The FAA (Milwaukee) then came to MSN and started an incident report. The aviation maintenance inspector said that since we flew around so long (37 minutes) investigating the problem, that we drew some suspicion. Come to find out from MSN tower, their FSDO (flight service district office) had instructed them to call any time an aircraft has a maintenance problem causing a return to the gate.

Report synopsis

Commuter, medium transport, got false warning on master warning panel after takeoff from MSN. PIC elected to return to MSN instead of proceeding to STL as recommended by maintenance.

Aircraft involvement

Aircraft was involved in an anomaly (an unsafe or illegal event).

Crew Starts Engines and Flies to Destination Even Though an Engine Exceeded Temperature Limits during First Start Attempt (ASRS Report #262820)

Date:	1994/01
Type of operation:	Air carrier, passenger flight
Aircraft:	Boeing B-727, low-wing, wide-body transport, between 150,001 and 300,000 lb
	Engines: Three turbojets
	Number of crew: Three
Flight crew:	ASRS reporter: captain, pilot-in-command. 15,700 hours, 145 hours within the past 90 days, 2100 hours in make/model.
	Co-ASRS reporter: second officer (S/O). 4500 hours, 240 hours within the past days, 3000 hours in make/model. Flight engineer rating, commercial license, instrument rating.

Narrative (unedited, ASRS reporter's own words)

During engine start, engine exceeded EGT (engine exhaust gas temperature) start limits. After consulting the B-727 flight handbook for guidance and our flight operations policy manual, it was determined there was nothing to preclude our trying a second start attempt.

Not Checking with Maintenance 163

Second start was successful, with start parameters well within limits.

Aircraft was then flown to STL (St. Louis) with all engine instruments well within normal operating limits. A logbook entry was made regarding the overtemperature during start of the #1 engine. The intent of this entry was to notify maintenance so they could perform their required check on the engine. Our flight handbook for this aircraft does not state this should or should not be a grounding item or guidance relating to this situation.

Supplemental information: #2 engine started first due to frozen start valve (normal start). #1 engine slow to rotate. Then when start lever was raised, the JT8D-15A stabilized at 520 degrees C. The temperature (EGT) then shot up to 660 degrees C approximately. All handbook procedures were complied with and engine was restarted later with no abnormal indications at all. Flight continued to STL. We requested that maintenance visually inspect engine. Their handbook told them to remove engine.

Supplement information: Upon arrival to aircraft, was advised by maintenance that previous aircraft laying over in these conditions could not start aircraft due to start valves being frozen and suggested to us to motor the starters. We completed the preflight and accomplished the before starting engine checklist. Then we motored #1 starter. Normal rotation. Then we motored #2 starter. Normal rotation. Then we motored #2 starter. No rotation. Then we motored #3 starter. Normal rotation.

We then called maintenance and accomplished a manual opening of the #2 starter valve. The indications were normal and accomplished a normal start. However, we noted a filter bypass annunciator illuminated. Then we motored #1 starter, all indications normal, raised the start lever. All indications normal until EGT rose past start limit, followed company procedures, motored starter for cooling and accomplished normal #1 start.

Report synopsis

B-727 flight crew experienced a hot engine #1 start at the ramp in cold weather. Aircraft engine was damaged. Crew started it again and continued to flight destination.

Aircraft involvement

Aircraft was involved in an anomaly (an unsafe or illegal flight).

Pilot Departs Airport with Multiple Aircraft Faults Displayed on Warning Panels (ASRS Report #425856)

Date:	1999/01
Type of operation:	Air carrier, passenger flight
Aircraft:	McDonnell Douglas MD 90, low-wing, medium-large transport, between 60,001 and 150,000 lb
	Engines: Two turbojets
	Number of Crew: Two
Flight crew:	ASRS reporter: first officer (F/O). Air transport certificate.

Narrative (unedited, ASRS reporter's own words)

During pushback, R engine system fail, R reverser fault illuminated, and on L side engine, generator would not come on line and a L AC fault illuminated. All 3 items require maintenance to clear and/or defer the item. (Engine system fail is not deferrable). The captain elected to take off and fly to maintenance facility in SJC (San Jose, CA). I don't agree with this but was overruled.

Not Checking with Maintenance 165

Report synopsis
An MD 90 PIC elects to depart SNA (Orange County, Calif.) airport with multiple aircraft faults displayed on the warning panels during pushback. F/O (first officer) was overruled in his desire to consult maintenance.

Aircraft involvement
Aircraft was involved in an anomaly (an unsafe or illegal event).

Crew Performed Takeoff with Fault Display Unit Showing No Information (ASRS Report #425920)

Date:	1999/01
Type of operation:	Air carrier, passenger flight
Aircraft:	Fokker FK10, low-wing, medium-large transport, between 60,001 and 150,000 lb
	Engines: Two turbojets
	Number of crew: Two
Flight crew:	ASRS reporter: captain, pilot-in-command. Air transport certificate.

Narrative (unedited, ASRS reporter's own words)
TUL (Tulsa, OK), Runway 36R. After gate departure, ATC (air traffic control) issued us a flow control time to ORD (O'Hare, Chicago, IL). Taxied to a hold area on 1 engine to conserve fuel. Started #2 engine and began the before takeoff checklist in time to make the flow

time. As we were cleared onto the runway to hold, we had difficulty getting the normal indication on the left MFDU (multiple fault display unit) during the takeoff configuration check. Verified all configuration systems were normal with no malfunctions indicated. Pushed the power up above minimum takeoff and did not get any takeoff configuration warnings.

With our flow control time about to expire and an aircraft on final for landing, I elected to take off. Enroute [sic] to ORD, I researched the problem and determined that I probably should have returned to the gate and had maintenance check out the system. Landed in ORD without incident. Did additional takeoff configuration checks during the taxi in at ORD and could not get the normal indication to appear on the L MFDU. Placed the discrepancy in the logbook.

Report synopsis
An FK10 flight crew performed a takeoff without their left MFDU displaying any information.

Aircraft involvement
Aircraft was involved in an anomaly (an unsafe or illegal event).

Recurring Gear Extension Problems Not Entered in Logbook Because of Assumption of Ice Sticking to the Gear (ASRS Report #426413)

Date: 1999/01

Type of operation: Commuter air carrier, passenger flight

Not Checking with Maintenance 167

Aircraft: Low-wing, light transport, between 14,501 and 30,000 lb
Engines: Two turboprops
Number of crew: Two

Flight crew: ASRS reporter: first officer (F/O). 1950 hours, 350 hours in the past 90 days, 350 hours in make/model. Air transport certificate.

Narrative (unedited, ASRS reporter's own words)

Flight took off from SLC (Salt Lake City) approximately XA30 January XA/99. We taxied through ice, snow and water. We took off and entered IMC (instrument meteorological conditions) about 1200 feet AGL (above ground level). We climbed to 2400 ft. with VMC (visual meteorological conditions). We then started the ILS (instrument landing system) into runway 25 at TWF (Twin Falls, ID). At final approach in VMC, about 1500 ft., I called for gear down before landing checks.

At that time, all red lights appeared and no gear down indications (including no gear down opening noise). New hire jump seat rider pulled out QRH (quick reference handbook) and read gear procedures for us. Captain then recycled gear 3–6 times. The gear then came down and locked. Captain then advised flight attendant that the gear came down and locked, and we would return for landing. Captain had advised ATC (air traffic control) we were in VMC and broke off the approach and would track LOC (ILS localizer) to airport. Tower advised us to stay at 6000–8000 ft. in VMC, and I did just that.

After that, the tower asked us how many souls and fuel on board. Captain answered. After the gear was down, Captain asked for the procedure turn and ILS approach again. We left the gear down. Did the approach and broke out at 1000 ft. AGL (above ground level). We were surprised to see fire truck at runway. We landed uneventfully. I stayed on aircraft to observe fueling and noticed lots of ice on gear during walka-round. (Whether it was there from SLC or during approach to TWF is unknown). Captain asked customer service to board, and the passengers then came out.

Captain returned and I inquired as to what maintenance control had said. He informed me that we were going to deice the wheel wells really well and also the aircraft and depart. We returned to SLC, and during an approach in IMC, again the same gear indications happened, but it only took 2 times to recycle gear. We landed uneventfully. Captain wrote up aircraft and maintenance took aircraft. During gear swings, maintenance found no mechanical problems. I later found out that the captain had been put on hold and hung up from maintenance control while in TWF. He had had the problem before many times (about 10) with ice and knew by his experience that is what they would have advised.

I feel that 2 things should have prevented this from becoming even an issue. First, the captain should have talked to maintenance control so that the problem back in SLC, when it happened again, could be shared with them instead of all the decision to depart TWF being the captain's. And second, the captain should have advised the F/O (first officer) and jump seat (rider) that he "never talked to maintenance control at the time of the departure" instead of leading me to believe that he had and that the deicing of the wheel wells was their idea, not his. Using CRM (crew resource management) and being honest with me, would have had myself suggest he get the mechanic's OK before departing.

Although he had the problem before, and he was relying on his experience, it's always a good idea to have mechanical support in case it turned out to be a mechanical problem and not ice, and informing the F/O the truth at TWF would have prevented the whole issue resting on the captain's shoulder, because I would have suggested he go call again. Then if he made a choice to leave, then it would be all his choice but maybe with him letting the rest of the flight crew believe it was a maintenance choice. This could become a problem for him.

Report synopsis
An EMB120 first officer reported on a recurring gear extension problem associated with slush and icing problems. The captain in this incident departed TWF without entering the gear problem in the logbook because of an assumption of ice sticking to the gear.

Aircraft involvement
Aircraft was involved in an anomaly (an unsafe or illegal event).

Oxygen Service Door Was Left Open Causing Flight Control Difficulties and Emergency Landing (ASRS Report #427215)

Date:	1999/01
Type of operation:	Commuter air carrier, passenger flight

Chapter 8

Aircraft: Canadian CL65, medium transport, between 30,001 and 60,000 lb

Engines: Two turbojets

Number of crew: Two

Flight crew: ASRS reporters: Captain, pilot-in-command. 4417 hours, 65 within the past 90 days, 37 hours in make/model. Air transport certificate. First officer (F/O). Commercial license, instrument rating.

Narrative (unedited, ASRS reporter's own words)

Crew waited for late inbound aircraft. As aircraft parked at the gate, maintenance began a service check. After maintenance left, we boarded and performed preflight checks. Maintenance returned to aircraft and kicked us out of aircraft cockpit to download aircraft data and perform minor repair on main cabin door. 10 minutes later, maintenance finished the door repair and returned logbook. We boarded passengers and taxied to runway 30. During the takeoff roll at 100 kts. a comparator warning message appeared along with F/O's airspeed reading approximately 40 kts. low compared to captain's and standby instruments.

Takeoff was continued due to runway surface conditions, and at about 160 kts., a loud buzzing sound was audible followed by the stall fail and mach trim fail EICAS messages. QRH (quick reference handbook) procedures were performed, and the aircraft returned for landing. Crew should not have assumed that only the door was worked on by maintenance, because another

Not Checking with Maintenance 171

mechanic, who serviced the oxygen while door work was being performed, left the oxygen service door open on the R side of the nose. We only inspected the cabin door before boarding. A thorough maintenance brief by maintenance or logbook notation would have alerted us to check access panels. A heads-up by maintenance that any work was going to be performed between crews would have been better, as we would not have accepted the aircraft until they finished.

Report synopsis
Oxygen service door was left open. It caused CL65 flight control difficulties on takeoff and emergency landing at IAD (Washington, Dulles Airport).

Aircraft involvement
Aircraft was involved in an anomaly (an unsafe or illegal event).

Freighter Dispatched with Left Navigation Light Inoperative in Darkness (ASRS Report #427290)

Date:	1999/01
Type of operation:	Air carrier, freight
Aircraft:	McDonnell Douglas DC8-73F freighter, low-wing, large-transport, between 150,001 and 300,000 lb
	Engines: Four turbojets
	Number of crew: Three
Flight crew:	ASRS reporter: first officer (F/O). 15,000 hours, 150 hours

within the past 90 days,
2600 hours in make/model.
Air transport certificate.

Narrative (unedited, ASRS reporter's own words)

While beginning taxi out of company owned ramp, ramp control coordinator advised us on company frequency that our L (red) position light was not illuminated. The R (green) and tail (white) lights were on. I asked the captain if he had heard the transmission to which he acknowledged, "Yes." I then asked what he wanted to do about this mechanical. He replied, "We'll take care of it later." After receiving taxi clearance from ground control, we completed the first stage of our normal taxi checklist. Upon completion, I asked the captain if he should contact maintenance for a deferral or if we should check the MEL. (Both of these items were company policy). Again, he replied, "We'll take care of it later."

After starting the remaining two engines, and upon completion of the entire taxi checklist, the F/E (flight engineer) queried the captain about checking the MEL or if he should enter the discrepancy in the logbook. Again, the captain deferred. We took off and flew SDF (Louisville, KY) direct MSP (Minneapolis/St. Paul) at night in violation of FAR 91 and in noncompliance of our own air carrier approved MEL and maintenance procedures. Our MEL does not allow us to fly at night without the position lights working. I was quite surprised that this captain disregarded procedures and the FAR, as my dealings with him in the past have always been very professional. Upon arrival at MSP, the discrepancy was entered in the logbook, and the position light relamped by maintenance.

Report synopsis
A DC8-73F was dispatched in noncompliance with the left navigation light inoperative in darkness in conflict with the MEL.

Aircraft involvement
Aircraft was involved in an anomaly (an unsafe or illegal event).

Summary and Assessment: Flight Crews Not Checking with Maintenance When Discrepancy Exists

This chapter includes 19 ASRS reports from 1988 through 1999 regarding flight crews not checking with maintenance when a discrepancy exists. The scenarios are all different, but the finale is the same. Some of the reasons for not checking with maintenance are as follows:

1. There is no resident maintenance at the airport (ASRS reports #92184, #90741, and #101677).

2. Checking with maintenance would result in a departure delay (ASRS reports #138343, #149395, #181587, #134560, #158584, and #92184).

3. Captain apparently accepts the risk, sometimes against the objections of the first or second officers (ASRS reports #138343, #427290, #262820, #182759, #181587, #134560, #425856, #425920, and #426413).

4. Pilot-in-command did not know that maintenance was required. PIC did not check logbook (ASRS reports #168110 and #114248).

These are classic flight crew human factors issues. Further discussion is provided in Chap. 10.

As indicated by these ASRS reports, operational situations are ripe with opportunities to justify breaking the rules in order to get the job done. Justification can be made for all sorts of violations, big and small. The psychological factor here is lack of self-discipline.

There are all sorts of reasons for a lack of self-discipline. There are also many organizational situations in which supervisory imposition of discipline is not easy or effective. There is only one solution to either case. *That solution is integrity*—a quality within that maintains the state of order and submission to rules and authority without the need for anyone besides the pilot him- or herself.

See Chap. 10 for further discussion.

9

Disagreement between Flight Crew and Maintenance and/or Management

In accordance with FAR part 91.7, the PIC (pilot-in-command) is responsible for determining whether an aircraft is in condition for safe flight. However, it is the responsibility of maintenance to perform all repairs and/or maintenance to provide a safe aircraft to the PIC. Then there is management, whose duty is to provide reliable transportation to the public and make a profit for the company.

Maintenance Inspection Waived to Avoid Delay (ASRS Report #91136)

Date: 1988/07
Type of operation: Air carrier, passenger flight
Aircraft: High-wing, light transport, between 14,501 and 30,000 lb
Engines: Two turboprops
Number of crew: Two

Flight crew: ASRS reporter: captain, pilot-in-command. 6000 hours, 200 hours within the past 90 days, 2000 hours in make/model. Air transport certificate.

Narrative (unedited, ASRS reporter's own words)

I was PIC (pilot-in-command) of airlines flight from DTW (Detroit, MI) to MQT (Marquette, MI) to CMX (Hancock, MI) to MSP (Minneapolis, St. Paul) with return as flight from MSP to CMX to MQT. The maintenance preflight on the aircraft was due to expire at XX07 local time (at MQT). The last leg of our schedule (CMX to MQT) was not due to leave until XX35 local time. I informed dispatch that the aircraft was in need of a maintenance preflight. I was then instructed to talk to director of maintenance. He informed me that he felt that the "circumstances beyond the control of the company" clause in the maintenance operations specifications allowed him to delay the preflight, rather than delay the flight.

I told him that if he gave me instructions to that effect in writing, I would fly the airplane. He refused to do so. I then went to the director of operations for instructions. He intimated that if I did not fly the aircraft, someone else would. I then asked him for written instructions. He refused but assured me that verbal instructions were sufficient. I feel that clarification of these rules is in order and that pilots should receive written instructions in cases where rules are in dispute.

Callback conversation with the reporter revealed the following: Operations specifications of the air carrier require a maintenance inspection of the aircraft every 72 hours. When inspection is accomplished, maintenance enters in the logbook the time, and then it is the responsibility of the flight crew to add 72 hours to this time to indicate when the next inspection is due. MQT is a maintenance facility, but in this case, the aircraft was allowed to continue to the next station where the inspection was accomplished. Time to do the inspection was about 30 minutes. Reporter feels very strongly that a clarification of these rules is in order.

Report synopsis
Required maintenance inspection was waived so aircraft could continue without delay.

Aircraft involvement
Aircraft was involved in an anomaly (an unsafe or illegal event).

Pilot Questions Controller's Clearing for Takeoff in Instrument Conditions with Transport on 3-Mile Final (ASRS Report #104394)

Date:	1989/02
Type of operation:	Commuter air carrier, passenger flight
Aircraft:	High-wing, light transport, between 14,501 and 30,000 lb
	Engines: Two turboprops
	Number of crew: Two
Flight crew:	ASRS reporter: first officer (F/O). 1850 hours, 120 hours

within the past 90 days, 450 hours in make/model. Commercial license, instrument rating.

Narrative (unedited, ASRS reporter's own words)

On a previous leg this morning, the captain and I noticed a higher than normal ITT (target torque) on the right engine during climb out. Also, at cruise, the engine became temperature limited at a lower altitude than we would have expected. We reported this to our maintenance facility while enroute [sic] to our destination. They considered the matter and gave it a deferred status. We continued the leg to our destination, landed and took off for our next destination and landed uneventfully. Leaving PIT (Pittsburgh, PA) we were cleared for an immediate takeoff. There was an aircraft on a 3 mile final at the time. Before V_1, I called, "Power set left, not set right." The captain attempted to correct the situation by advancing the right power lever and elected to continue the takeoff.

We reached V_1 and V_R and rotated and climbed out. We experienced conditions similar to the ones experienced with the right engine while enroute. The captain elected to divert to MDT (Harrisburg) based on the weather and conditions at our destination and his lack of confidence in the right engine should we need to execute a missed approach. Maintenance did a ground run up of the aircraft. They found nothing wrong and returned the aircraft to service. Although this cannot be considered a real incident, several things came to mind. Was maintenance correct in telling us to continue once we reported our first findings? Was the captain justified in continuing the takeoff in PIT (Pittsburgh) although target torque was not achieved on the right engine? Did maintenance do a thorough enough check of the aircraft after the diversion? I believe the captain felt pressure to continue the takeoff because of the weather and the aircraft on final and possibly his previous experience with maintenance telling him things were acceptable although he believed otherwise.

Supplemental information: At V_1/V_R speed, I noticed a slight mechanical problem that should have called for an aborted takeoff. Why would the tower clear any aircraft to takeoff and another to land on the same runway in such low visibility conditions when they could not see me nor the other aircraft?

Report synopsis
PIC (pilot-in-command) of commuter light transport questioned local controller's technique for clearing him for takeoff in IMC (instrument meteorological conditions) with a medium-large transport on a 3-mi final. First officer is more concerned about aircraft maintenance procedure discrepancy.

Aircraft involvement
Aircraft was involved in an anomaly (an unsafe or illegal event).

MEL Interpretation Dispute between Flight Crew and Maintenance (ASRS Report #108983)

Date: 1989/04
Type of operation: Air carrier, passenger flight

Chapter 9

Aircraft: Low-wing, large transport, between 150,001 and 300,000 lb
Engines: Three turbojets
Number of crew: Three

Flight crew: ASRS reporter: second officer (S/O), flight engineer. 5600 hours, 200 hours within the past 90 days, 2500 hours in make/model. Air transport certificate, flight engineer rating, certified flight instructor.

Narrative (unedited, ASRS reporter's own words)

Upon arrival at the aircraft, air carrier flight XX from DEN (Denver, CO) to OAK (Oakland, CA), I completed my SOP (standard operating procedure) cockpit set up. The aircraft, XXXX, had arrived at the gate in DEN about 5 minutes before I got there. I proceeded to do my walkaround exterior inspection and noticed that there was a total of 2 broken static bonding wires on the #3 (left inboard) and #6 (right outboard) leading-edge flaps. I immediately reported this discrepancy to DEN line maintenance. A few minutes later, our mechanic and supervisor came to the cockpit and demanded to see where these broken bonding wires were. I promptly showed them. I knew that it was required to operate the large transport with at least one bonding wire per leading edge device.

The maintenance supervisor read the air carrier's maintenance manual and decided to clip the broken bonding wires off. His interpretation of the manual was that only one bonding wire need be functional for each side of the aircraft. I questioned him on this, and he assured me that it was OK to go this way. He seemed in a big hurry due to the fact that our scheduled departure time was rapidly approaching. At this air carrier, a maintenance person's performance is judged by how close to departure time they can get a flight out. If a certain mechanic has too many delays, he is not promoted as quickly. I find this policy to be a safety hazard.

We left the gate on time and proceeded to OAK. Upon arrival, I did another walkaround and found that the maintenance supervisor had, indeed, clipped the bond wires on #3 and #6. I called our maintenance controller in SFO (San Francisco), and he told me we were required one bond per leading edge device. Needless to say, we got them replaced in OAK and were late on returning flight. The fact that pilots of our air carrier airlines are not provided with aircraft MELs (minimum equipment list) is definitely unsafe. I have to trust a mechanic to interpret his maintenance manual correctly.

We were dispatched with an illegal aircraft, and yet we were assured by our maintenance people that it was legal. I need the information to make up my own mind—after all, it is my FAA license riding on the outcome. Most other FAR 121 air carriers have the MEL in the cockpit of every one of their aircraft. Why it is not required of our air carrier is beyond me. This information is known only to our main maintenance people in SFO. I realize that maintenance delays are not part of the public reporting system that the airlines use; however, our air carrier has some sort of program to encourage mechanics to do jobs quickly. I'd personally rather have the job done correctly and legally.

Report synopsis

There was a minimum equipment list interpretation dispute between a flight crew member and the maintenance supervisor regarding the legality of dispatch with broken static bonding wires.

Aircraft involvement
Aircraft was involved in an anomaly (an unsafe or illegal event).

Captains of Two Transports Question Legality of Flying Aircraft with Electric Trim Inoperable (ASRS Report #109052)

Date:	1989/04
Type of operation:	Commuter air carrier, passenger flight
Aircraft:	Low-wing, light-transport, between 14,501 and 30,000 lb
	Engines: Two turboprops
	Number of crew: Two
Flight crew:	ASRS reporter: captain, pilot-in-command. 9000 hours, 250 hours within the past 90 days, 250 hours in make/model. Air transport certificate.

Narrative (unedited, ASRS reporter's own words)
Enroute [sic] from SCK (Sacramento, CA) to SFO (San Francisco, CA), the electric trim system was determined to be inoperative. The aircraft was trimmed

Crew, Maintenance, Management Disagreements 185

manually, and the flight continued without incident. The autopilot was selected on enroute, and it was noted that the altitude hold function of the autopilot was inoperative. Upon landing at SFO, maintenance control was notified, and a mechanic was dispatched to the aircraft. The mechanic determined that the autopilot elevator servo was inoperative and was the cause for both the electric trim and autopilot not to function by manufacturer's design.

The mechanics advised that the corrective action for this discrepancy would be to defer the write up per the MEL as directed by maintenance control. The deferral being applied to the discrepancy was pertaining to the autopilot system. The MEL clearly states that any one function of the autopilot may be inoperative, that remaining operating functions may be used. I disagreed with this corrective action. The discrepancy was specific to the electric trim being inoperative, not the autopilot system. The electric trim uses a component of the autopilot system to operate. The electric trim is independent of the autopilot system for it to function as an electric trim system. The electric trim system is not listed in the MEL; therefore, the electric trim must be considered an integral part of the aircraft and not considered deferrable.

I related my concern and disagreement with the corrective action to the on-site mechanic, then, via phone, to the maintenance control on duty and finally, the management pilot on duty who was, at the time, the chief pilot. The chief pilot explained that maintenance control had researched this problem and determined, in fact, that the electric trim could be deferred under this MEL reference. I further explained to the chief pilot my concern over the legality of operating under this MEL reference and related to him a recent incident concerning the improper use of the MEL by maintenance control. The chief pilot assured me that maintenance control would be responsible should the FAA question my authority in accepting the aircraft for return to service under these conditions. The chief pilot advised me to fly the aircraft. I did not believe there was any safety of flight issue re: the airworthiness of the aircraft with the autopilot deferral under these conditions. The assumed liability and authority was taken by the company when the chief pilot advised me to fly the aircraft.

Supplemental information: I was assigned an aircraft that had the electric trim deferred. I was of the opinion that this item had been deferred in error. I contacted maintenance control to discuss the situation. Shortly after, I received a phone call from dispatch. They wanted to know what my problem was and directed me to call the chief pilot at home. I phoned the chief pilot at home, and he, too, wanted to know what my problem was. I told him I was of the opinion the electric trim had not been deferred correctly and that, to the best of my knowledge, it was a violation of the FARs to fly the aircraft. He told me it was not my responsibility to determine the legality of the deferral. By the fact that maintenance control had deferred it absolved me of all responsibility for flying the aircraft in this condition.

I indicated my dissatisfaction with this logic but to no avail. He ordered me to fly the aircraft under threat of termination. The airline's management regularly "interprets" the MEL in order to keep aircraft flying. This includes the scenarios as above and avoiding repairs by allowing open deferrals to exist for weeks at a time. Our flight group is constantly reminded by management, verbally and in writing, that our go/no go decisions as PICs are subject to review, suspension and termination.

Report synopsis
In two reports captains of light transport questioned the Federal Aviation Regulations legality of flying the aircraft with electric trim inoperable. Flight was approved by maintenance per MEL (minimum equipment list).

Aircraft involvement
Aircraft was involved in an anomaly (an unsafe or illegal event).

Captain's Complaint about Poor Company Maintenance (ASRS Report #124262)

Date:	1989/10
Type of operation:	Commuter air carrier, passenger flight
Aircraft:	Low-wing, transport, between 14,501 and 30,000 lb
	Engines: Two turboprops
	Number of crew: Two
Flight crew:	ASRS reporter: captain, pilot-in-command. 2700 hours, 300 hours within the past 90 days, 1500 hours in make/model. Air transport certificate, certified flight instructor.

Narrative (unedited, ASRS reporter's own words)

MEL calls for 1 of 2 air cycle machines to be working with the restriction that if 1 is inoperative, other must provide for adequate heating and cooling for normal operations. Picked up aircraft at home maintenance base, and the left air conditioning group was inoperative. Flew one leg and squawked other air conditioning group as not providing adequate cooling. Since we were at an outstation, maintenance control elected to defer the pressurization system so that we could fly the aircraft back to a maintenance base. I argued about this, believing that it didn't address the problem, which had been recurring.

Finally maintenance convinced me to fly the aircraft. Contributing to my decision were problems that we, as a company, have been having staying in business. I didn't want to lose a full load of passengers and the associated revenue. Also feared some retaliation from management if the plane was grounded at an outstation. After repeated squawks concerning the same problem on this aircraft and others, I have become convinced that the FAA does not monitor our maintenance enough, and the FAA doesn't want to make waves. I've even called our maintenance manager to ask him to persist in his efforts to make these air conditioning [units] safe, but he's never around. Feel we've been abandoned.

Report synopsis

Commuter captain complained about poor company maintenance.

Aircraft involvement

Aircraft was involved in an anomaly (an unsafe or illegal event).

Transport Failed to Get Required 48-Hour Maintenance Check Prior to Departure (ASRS Report #124945)

Date: 1989/10
Type of operation: Commuter air carrier, passenger flight
Aircraft: Low-wing, light transport, between 14,501 and 30,000 lb
Engines: Two turboprops
Number of crew: Two
Flight crew: ASRS reporter: first officer (F/O). 4500 hours, 280 hours within the past 90 days, 430 hours in make/model. Air transport certificate, certified flight instructor.

Narrative (unedited, ASRS reporter's own words)

Aircraft requires an "A" check every 48 hours as per company operations with FAA. Aircraft was fogged in the previous night, so it did not make it back to SFO (San Francisco, CA) so this check could be performed. An "A" check consists of all lights, fluids, brakes, tires, etc., being checked by a mechanic. After talking to maintenance control, we were told and assured that approval had been obtained from the FAA to extend this check an additional 5 hours.

We then called the management pilot on duty and had him check on this, and he assured us that this extension had been approved and obtained. After making the flight to SFO, we discovered that both maintenance and the management pilot on duty had told us this so the company could get the aircraft to SFO. Required maintenance checks cannot be extended.

Report synopsis
Commuter light transport failed to get required 48-hour maintenance check prior to departure on scheduled flight.

Aircraft involvement
Aircraft was involved in an anomaly (an unsafe or illegal event).

Owner and Pilot Performed Minor Maintenance (ASRS Report #130296)

Date:	1989/11
Type of operation:	Training flight in small aircraft
Aircraft:	High-wing, fixed-gear, small aircraft, less than 5000 lb
	Engines: One reciprocating
	Number of crew: One
Flight crew:	General aviation, single pilot. 800 hours, 100 hours within the past 90 days, 600 hours in make/model. Commercial license, instrument rating.

Narrative (unedited, ASRS reporter's own words)
I borrowed a friend's small aircraft to take my CFI (commercial flight instructor) checkride at Baltimore

Crew, Maintenance, Management Disagreements 191

FAA FSDO (Flight Service District Office) 7 on Nov/Thu/89. The safety inspector gave the aircraft's log to a maintenance inspector for review. The maintenance inspector returned to me with the plane's log and said the plane was illegal to fly due to owner performed maintenance. He itemized: 1) Aug 1989: The owner replaced a leaking wet compass with a card compass. 2) August 1989: The owner replaced the facility avionics fan with an identical manufacturer supplied fan and blower. 3) November 1989: I recalculated the weight and balance to reflect the removal of main wheel pants for winter operations.

Neither the owner nor I are [sic] A&Ps. An A&P did not sign off the owner's work nor my recalculations. The maintenance inspector maintained I used an incorrect arm to redo the weight and balance. Additionally, the owner and I swung the compass, not once, but twice (10 days apart), to confirm the deviation card. Upon being informed by the maintenance inspector's comments, the owner immediately took the plane to the avionics shop at Baltimore before the flight check part of my exam. His intent was to comply as soon as possible, as he and I were unaware of these maintenance violations. We were unaware, because we had consulted with our regular mechanic about doing the work, and the mechanic said it would be legal for us to do this work. When the work was inspected at Baltimore, all work was signed off without any changes, including the deviation card.

Supplemental information: I replaced the avionics cooling motor and fan assembly with an approved motor and fan assembly from manufacturer aircraft corporation. This fan and motor assembly is attached by 4 screws to the firewall of the aircraft and uses 2 quick connect electrical connections. I replaced a defective wet type compass with a TSO'ed* company vertical card compass. The compass was mounted on the existing support used for the wet compass per the manufacturer's directions. I temporarily removed my main landing gear wheel pants. Subsequently, the weight and balance were recalculated. I believed at the time this work was accomplished under owner allowed maintenance and confirmed in part by my regular A&P mechanic.

Report synopsis
Owner and pilot friend performed minor maintenance on aircraft.

Aircraft involvement
Aircraft was involved in an anomaly (an unsafe or illegal event).

After Engine Start Fire, Cabin Attendant Deployed Emergency Exit without Pilot's Instruction (ASRS Report #138859)

Date:	1990/03
Type of operation:	Air carrier, passenger flight
Aircraft:	Low-wing, medium-large transport, between 60,001 and 150,000 lbs
	Engines: Two turbojets
	Number of crew: Two

---

*TSO = Technical Standard Order. FAA technical requirements for the design and manufacture of aircraft instruments and equipment. Approved items are considered "TSOed."

Crew, Maintenance, Management Disagreements

Flight Crew: ASRS reporter: first officer (F/O). 7000 hours, 250 hours within the past 90 days, 2500 hours in make/model. Air transport certificate.

Narrative (unedited, ASRS reporter's own words)

I was performing the duties of F/O (first officer) on a medium large aircraft. After pushing back from the gate, we could not get #2 engine to light off. We returned to the gate, advised maintenance. After approximately 1½ hours, maintenance replaced an ignition relay system, got the engine to start, signed off the logbook and said it was fixed. We pushed back from the gate, tried to start #2 engine and could not get it to light off. Again, we advised maintenance. We started the #1 engine. Then maintenance requested that we try to start the #2 engine again by a nonstandard procedure. The procedure was emergency power switch to "on," ignition switch to "override." We advised that this was not in our flight manual or an approved procedure for starting an engine at the gate. They acknowledged that fact but requested we do that to assist them in troubleshooting. We did what was requested of us and got a hot start on the #2 engine.

The pushback crew advised us of smoke and a flame in the tail pipe of the #2 engine. After securing the #2 engine from the attempted start, the ground crew said it was still smoking profusely and still had intermittent flames. At that time, we contacted the senior F/A (flight attendant) and told him we had better get the passengers off the airplane. He asked if he should order an evacuation via the slides. We told him that would not be necessary, to just use the front entry stairs. In the ensuing confusion, an F/A in the aft part of the airplane jettisoned the tail cone. However, the tail slide did not inflate as advertised. The aircraft was deplaned using the stairway. To prevent a recurrence of this situation, never try to accomplish maintenance procedure, and F/As should not jettison emergency exits until they are told to via the FAA approved emergency evacuation plan as commanded from the cockpit or senior F/A over the PA system.

Report synopsis
Flight crew of air carrier, medium-large aircraft, had trouble starting the engine. After maintenance activity and repeated start attempts, they got a stack fire, and cabin attendant deployed an emergency exit without PIC (pilot-in-command) instructions.

Aircraft involvement
Aircraft was involved in an anomaly (an unsafe or illegal event).

Flight Crew and Maintenance Disagree on Interpretation of MEL Requirements and Logbook Sign-Off (ASRS Report #154204)

Date: 1990/08
Type of operation: Air carrier, passenger flight
Aircraft: Low-wing, large transport, between 150,001 and 300,000 lb

Engines: Three turbojets

Number of crew: Three

Crew, Maintenance, Management Disagreements

Flight crew: ASRS reporter: second officer (S/O) (flight engineer). 4000 hours, 30 hours within the past 90 days, 30 hours in make/model. Air transport certificate, flight engineer rating.

Narrative (unedited, ASRS reporter's own words)

A preflight examination of the aircraft logbook showed a placard which stated that both main landing gear door hinge covers were removed/missing for maintenance. The preflight walk around revealed that only one hinge cover door was missing on each side. (The forward ones were missing). The aircraft configuration deviation list (CDL) shows that there are 2 hinge covers on each gear and states that 2 on one side may be missing. There is no mention in the CDL about one missing hinge cover on each side.

The captain requested clarification/approval to launch with a missing hinge cover. Maintenance directed that it was OK to fly. The captain required maintenance to write it up that way in the logbook, then we launched. Maintenance personnel verified from aircraft manufacturer that the aircraft was not certified to fly in that configuration. The aircraft had been operated in that configuration for several days prior to our flight.

Supplemental information: I called maintenance at DFW (Dallas/Ft. Worth, TX) plus tech service for aircraft type and was told in rather strong terms that the sign off was legal, and the aircraft agreed with CDL restrictions. I found out I was right, and maintenance was wrong, and I was wrong in taking off. Aircraft not airworthy.

Airline management followed up this resulting delay (20 minutes) with an unsigned company radio message indicating I was charged with a crew delay, "debrief on hotline." This message sent during sterile cockpit period on taxi out. Receipt of this administrative admonishment prior to takeoff, in my opinion, directly affected the safe operation of the flight due to the mental frustration of the events.

Report synopsis
Air carrier, large transport, flight crew and maintenance personnel disagree on interpretation of MEL requirements and subsequent logbook sign-off. Flight crew took flight. However, maintenance subsequently discovered that flight was in violation of MEL and manufacturer's recommendations. Flight dispatch issued an admonishment via company radio about the delay, during a sterile cockpit period.

Aircraft involvement
Aircraft was involved in an anomaly (an unsafe or illegal event).

Captain Coerced into Accepting Aircraft with Door Latch Problems (ASRS Report #156013)

Date:	1990/09
Type of operation:	Air carrier, passenger flight
Aircraft:	High-wing, medium transport, between 30,001 and 60,000 lb
	Engines: Four turboprops
	Number of crew: Two

Crew, Maintenance, Management Disagreements

Flight crew: ASRS reporter: captain, pilot-in-command. 8000 hours, 215 hours within the past 90 days, 3000 hours in make/model. Air transport certificate, flight engineer rating.

Narrative (unedited, ASRS reporter's own words)

While waiting for aircraft to come in from PHL (Philadelphia, PA), talked with dispatch concerning aircraft mechanical problems (we were late and another crew was bringing it in from PHL). I was told that the problem was with the door seal (pressurized aircraft). When the aircraft arrived in DCA (Ronald Reagan National, Washington, DC), I talked with the crew and found out that it was the door closing mechanism. A lever is pushed down over center, which inflates the door seal and keeps the door immobile. (The door is heavy, and this same lever, when lifted, raises the whole door over lips so that it can open outward).

The other crew told me that the door was fixed with no problems on their flight but that the lever was "spongy." It was, but all door lights went out when it was closed, so I accepted the aircraft. Then I discovered that the pressurization had been MEL'd (minimum equipment list). Although the door seal (pressurization) is an integral part of the lever, the lever itself was what had been giving problems.

I took the aircraft as previous crew had said "no problem." Enroute [*sic*], the handle popped up just over center, and the door seal deflated. A passenger asked the F/A (flight attendant) if she needed help holding it down (she said no and left the lever as it was. Door did not open).

My main concern is that in my opinion (after the fact), I feel that the aircraft should not have been in revenue service. I also feel that the company had a lot more facts than I did. They maneuvered me and my crew into operating a questionably airworthy aircraft.

Report synopsis
Captain of air carrier, medium transport, complained about being coerced into accepting an aircraft with a maintenance listing of door latch problems. Door latch malfunctioned in flight.

Aircraft involvement
Aircraft was involved in an anomaly (an unsafe or illegal event).

Captain Refused Aircraft after Maintenance Examined Stabilizer, but Reconsidered after Chief Pilot Examined and Accepted Aircraft (ASRS Report #163234)

Date:	1990/11
Type of operation:	Air carrier, passenger flight
Aircraft:	Low-wing, large transport, between 150,001 and 300,000 lb
	Engines: Three turbojets
	Number of crew: Three

Flight crew: ASRS reporter: captain, pilot-in-command. 12,000 hours, 150 hours within the past 90 days, 2000 hours in make/model. Air transport certificate, flight engineer rating, certified flight instructor.

Narrative (unedited, ASRS reporter's own words)

On arrival at aircraft, found maintenance clearing a previous item. Inbound crew wrote up main stabilizer trim as making a clicking noise. Maintenance advised they had inspected aircraft and found airworthy. We reviewed (maintenance and I) stabilizer trim. Trim clicking noise was still found. Informed maintenance did not like the aircraft and wanted it fixed. Their response was that they considered it OK and was airworthy. Called chief pilot's office and two chiefs were dispatched to aircraft. Chiefs conferred with maintenance and found aircraft acceptable. They also concurred with noise heard in stabilizer trim.

Asked if chiefs would take aircraft. They said yes. At this point, agreed to take aircraft. Flew two legs and clicking noise did not get worse or better. On arrival on second leg, wrote noise up again. It didn't feel like aircraft was fixed but didn't feel like it was unsafe either! I was put between a rock and a hard place. To complicate matters, contract negotiations were on going [sic] and company very sensitive to cockpit complaints. I don't think I did anything wrong, but I could see how it could look that way; hence this report.

Report synopsis

Airline maintenance declared the aircraft airworthy after examining and considering the previous logbook write-up on the main stabilizer. The captain refused the aircraft. Chief pilots examined the aircraft and accepted the aircraft. The captain reconsidered and accepted the aircraft and flew his trip.

Aircraft involvement
Aircraft was involved in an anomaly (an unsafe or illegal event).

Flight Crew and Maintenance Disagree on Whether Logbook Write-up Permits Flight (ASRS Report #173054)

Date:	1991/03
Type of operation:	Air carrier, passenger flight
Aircraft:	Low-wing, wide-body transport, over 300,000 lb gross weight
	Engines: Three turbojets
	Number of crew: Three
Flight crew:	ASRS reporter: captain, pilot-in-command. 25,000 hours, 250 hours within the past 90 days, 2500 hours in make/model. Air transport certificate.

Narrative (unedited, ASRS reporter's own words)
Inbound had one "inoperative" placard on the F/E (flight engineer) panel. It stated that #2 Hi-stage pneumatic supply was inoperative, use Lo-stage only. We

operated the flight JFK-LAX (John F. Kennedy, New York, NY–Los Angeles, CA) using #2 Lo-stage pneumatic. On descent, #2 pneumatic was turned off and written up in the logbook. Los Angeles maintenance noted that in reviewing the logbook that #2 pack temperature indicator was previously inoperative and that #2 pneumatic system should not be used. They feel that FAA logbook review may consider this as use of an inoperative system.

There were 3 inoperative and #2 pack inoperative in automatic. There were no placards that stated #2 pneumatic was unusable! We check the placards against our MEL and found no restrictions other than "use #2 Lo-stage" and "operate #2 pack in manual." On preflight, our crew reviewed the logbook and MEL. JFK maintenance also did same. If #2 pneumatic was not usable, it should have been placarded: "Do not pressurize, except engine start." If #2 pack was inoperative, it should have been placarded inoperative. The whole thing sounds like a matter of interpretation. We should not use an inoperative system. We would not be dumb enough to write up a malfunction of an inoperative system.

Report synopsis

Flight crew of wide-body aircraft interpreted logbook write-up as permitting flight. Upon reaching the destination, maintenance personnel considered the write-ups prohibitive for flight.

Note Pack is an air conditioning unit.

Aircraft involvement

Aircraft was involved in an anomaly (an unsafe or illegal event).

After Engine Exceeds Maximum Temperature Flight Crew Deviates from Airline Operational Procedure with Maintenance Encouragement (ASRS Report #183009)

Date: 1991/07
Type of operation: Air carrier, passenger flight
Aircraft: Heavy, low-wing transport, over 300,000 lb
Engines: Four turbojets
Number of crew: Three
Flight crew: ASRS reporter: captain, pilot-in-command. 14,000 hours, 180 hours in the past 90 days, 1100 hours in make/model. Air transport certificate.

ASRS reporter: second officer (S/O). 3200 hours, 130 hours within the past 90 days, 760 hours in make/model. Air transport certificate, flight engineer rating.

Narrative (unedited, ASRS reporter's own words)

At lift off, runway 8R, Honolulu, #4 engine overtemped (caused by failure of power management control unit). Thrust lever was immediately retarded and clean-up and climb continued. All engine parameters were normal. At approximately 10000 ft., EGT (engine exhaust gas temperature) recall pushed. #4 recorded at 956 degrees C. Duration of overtemp approximately 10–15 seconds. Referred to flight manual limits and 930 degrees C was never exceed temperature. Engine shut down. All parameters still normal. Contacted system line maintenance via ARINC (Aeronautical Radio, Inc.) and advised of situation.

Temperature recalled, duration and engine operating normal, asked if they wanted aircraft returned to HNL or continued to LAX (Los Angeles, CA). After some delay (minutes), the response was, "If the engine is operating in climb in normal parameters, and if you are comfortable with it, we would like the aircraft in LAX for engine change." Balance of segment uneventful. Aircraft was subsequently ferried by EC crew to SFO (San Francisco, CA) for "engine change." Perhaps the decision to continue to LAX was unduly influenced by the "engine experts" decision to continue after our best information was passed on.

Supplemental information: Immediately after takeoff, I noticed an overtemp of #4 EGT. N_1 N_2 were also abnormally high. I advised the captain, and he throttled #4 back to where the engine was matching the others. After climb out, we looked in the flight manual where it said the engine should be shut down due to the exceeded EGT value. In the conversation, I remember the captain say[ing] the "book" said to shut down the engine. But maintenance said to check all parameters, and if the engine is running OK, leave it running and continue. There was a discussion among the crew, and we concurred with maintenance's judgment. There now seemed to be some second guessing from others, the flight crew included, about not shutting down the engine. At the same time, we felt safe to continue and I still do.

Callback conversation with the reporter revealed the following information: The S/O (second officer, flight engineer) reporter felt that, in retrospect, the flight crew had erred in not shutting down the engine. The S/O also felt that had he mentioned the fact that the book said to shut the engine down and had inferred that it would be best to return to HNL, the captain may have complied. Reporter said there was a lack of communication in the cockpit on this issue and a feeling or thought of get homeitis or schedule pressure was not brought up in any conversation. There was a later meeting with the air carrier middle management where the flight manager discussed the situation and felt that maintenance had not offered the proper choice to the flight crew. Reporter feels that this will not happen to this crew a second time. The S/O reporter is in his 12th month of operation with air carrier.

Report synopsis
Flight crew deviation from airline operational procedure was exacerbated by airline maintenance policy procedures after engine exceeded maximum temperature for operation.

Aircraft involvement
Aircraft was involved in an anomaly (an unsafe or illegal event).

Transport Dispatched in Questionable Airworthiness Condition after Hail Damage Repair Accomplished (ASRS Report #426595)

Date: 1999/01
Type of operation: Air carrier, freight

Aircraft:	DA20 Falcon, low-wing, medium transport, between 30,000 and 60,000 lb
	Engines: Two turbojets
	Number of crew: Two
Flight crew:	ASRS reporter: captain, pilot-in-command. 3300 hours, 150 hours within the past 90 days, 900 hours in make/model. Air transport certificate, certified flight instructor.

Narrative (unedited, ASRS reporter's own words)

Aircraft received hail damage to radome, navigation lights, leading edges and engine inlets on descent into SHV (Shreveport, LA) on Jan/XA/99. Company maintenance personnel was flown in to fix the plane. I arrived later that day and was then assigned to that aircraft. On Jan/XC/99, I was assigned a trip out of SHV. The aircraft log was signed as airworthy. All squawks were cleared by our company maintenance personnel. While preflighting the plane, I noticed dimpling on the wing leading edges and on engine inlets and bullets. I checked the fans and front turbine blades and saw no obvious damage.

I proceeded to fly the trip assigned to me and ended at our company maintenance base in ZZZ. The next day, I'm told that the plane had "substantial damage" and should not have been flown. Most of our planes have dimples on the leading edges, but I am told that those are "within tolerances." How am I supposed to know that the dimples in the one I flew were not "within tolerances" and constitute "substantial damage"? I was counting on our professional maintenance personnel to certify that the plane was airworthy and they did.

Callback conversation with the reporter revealed the following information: The reporter stated the preflight check revealed some small dents in the wing and stabilizer leading edges but nothing that the rest of the fleet did not display. The reporter said the maintenance release was signed in the logbook, and the airplane was legal to fly. The reporter later was advised the wing leading-edge dents were out of limits, and the aircraft was deemed unairworthy. The reporter said the FAA was contacted by the reporter and he was absolved from any culpability of operating the aircraft in noncompliance.

Report synopsis
A DA20 Falcon was dispatched in a questionable airworthiness condition after hail damage repair was accomplished.

Aircraft involvement
Aircraft was involved in an anomaly (an unsafe or illegal event).

Summary and Assessment: Disagreements between Flight Crew and Maintenance or Management

In accordance with FAR part 91.7, the pilot-in-command is responsible for determining whether an aircraft is in condition for a safe flight. However, it is the responsibility of maintenance to perform all repairs and/or maintenance to provide a safe aircraft to the PIC. Then

Crew, Maintenance, Management Disagreements 207

there is management, whose objective is to provide reliable service to the public and to make a profit for the company.

This chapter includes 14 ASRS reports where the PIC is persuaded and even coerced by maintenance and management into accepting an aircraft against his better judgment. In one case, the PIC was even threatened with termination if he refused the aircraft.

The attempt to satisfy the demands of the aircraft operator and its maintenance department, as well as save his own career and ensure flight safety, sometimes puts the PIC in a compromising position, as indicated in these 14 ASRS reports.

1. Maintenance and management persuaded PIC to accept aircraft they said was airworthy, even though PIC believed it was not airworthy, just to get aircraft out on time or to a maintenance base. After the fact, it was proved that the PIC was right, but he flew an unairworthy aircraft (ASRS reports #91136, #108983, #109052, #124945, #154204, #173054, #183006, and #426595).

2. PIC threatened with termination if he did not accept aircraft he believed to be unairworthy (ASRS report #109052).

3. PIC continued flight of aircraft with air conditioning system problems out of fear of retaliation by management (ASRS report #124262).

4. Small aircraft owner and pilot friend performed maintenance with mechanic's verbal OK, but mechanic did not approve in writing. FAA said their work was illegal (ASRS report #130290).

5. Against PIC objections, maintenance persuaded PIC to perform unapproved engine start. Engine overtemperatured, resulting in start fire. Passengers were deplaned (ASRS report #138859).

6. PIC was charged with delay regarding dispute over MEL'd item even though PIC was right (ASRS report #154204).

7. PIC was coerced into accepting aircraft which should have been grounded due to door latch problems (ASRS report #156013).

8. Maintenance declared aircraft airworthy with stabilizer problems. PIC refused to accept aircraft, but chief pilot said aircraft was OK, so PIC reconsidered and accepted aircraft (ASRS report #163234).

9. Flight crew deviation from airline policy procedures is exacerbated by airline maintenance's advice after engine exceeds maximum temperature and continues flight from Honolulu to Los Angeles (ASRS report 183009).

Although each of these cases could be considered pilot error, it is hard to fault the pilot, confronted with such a formidable adversary such as maintenance backed by management.

See Chap. 10 for further discussion regarding these "pilot error" issues.

10

Summary and Final Discussion

Although a few ASRS reports included operators of small, reciprocating engine powered aircraft, the number is too small to arrive at any significant conclusion. The majority of pilot reporters flew turbine-powered FAR part 121 air carrier aircraft.

Pilot reporters were all qualified with air transport certificates and high flight times. Major air carrier captains' flight time exceeded 10,000 hours while commuter captains' experience was generally less than 10,000 hours. These air carrier pilots are all exposed to comprehensive training programs and flight and simulator checks.

The air carrier is in business to transport passengers and make money for the company and stockholders. Safety, per se, is not an objective. If the primary objective were safety, the only way to achieve this would be to lock the aircraft in the hangar and not fly them. In other words, the role of safety is in support of the operator's primary mission.

Whereas the aircraft operator's primary objective is not safety, safety is the primary role of the FAA, and the FAA has given the pilot-in-command the final decision regarding the safety (airworthiness) of the aircraft that the operator has provided for him to fly.

Between management and the flight crew is maintenance. As indicated by these 79 ASRS reports from 1988 through 1999, the interaction between maintenance and the flight crew is not always smooth. Attempting to satisfy the demands of the aircraft operator and its maintenance department and to ensure flight safety sometimes puts the pilot-in-command in a compromising position, leading to "pilot error." Much of this pilot error could be eliminated by the air carrier and its training program. Applying crew resource management (CRM) techniques—possibly even joint training classes—could improve pilot-mechanic interaction in maintenance discrepancy reporting.

About 85 percent of all pilots and maintenance technicians believe miscommunication occurs between them at least half the time when reporting or attempting to resolve maintenance log entries, according to an interdisciplinary survey conducted by students and faculty at Purdue University, including some from the school's Department of Aviation Technology. Reasons cited included lack of detail or inadequate time to carry out a debriefing between flights.

A standardized postflight debriefing or interview process requiring representatives from both specialties also would help, most respondents said. Although the study covered a broad range of operations, most participants flew or maintained turbine equipment.

Although training programs could eliminate many pilot error incidents, the pilot may be too dependent on them, expecting the training program to cover all eventualities.

Summary and Final Discussion 213

Of the eight pilot error categories covered in the preceding chapters, four stand out based on the proportionally high number of ASRS reports:

1. Problems with MEL: 17 reports.
2. Illegal or unsafe flight due to improper maintenance paperwork, which the PIC did not discover: 13 reports.
3. Flight crew not checking with maintenance when discrepancy exists: 19 reports.
4. Disagreement between flight crew and/or management: 14 reports.

Resolution of Error and Lessons Learned

Regarding problems with MEL, granted, specialized training would resolve many issues involving MEL. Besides, *maintenance error* is also a factor. Since we are concerned with pilot error, possibly the PIC needs to do some homework such as study the MEL, discuss it with maintenance, and above all, be familiar with the inevitable *flight restrictions* resulting from deferred components or systems.

If the PICs are thoroughly familiar with the MEL before they are confronted with it, then when pushed for time just before departure, for example, they are in a more favorable position to make the proper decisions regarding the flight.

Regarding item 2 above, illegal or unsafe flight due to improper maintenance paperwork is a maintenance error as well as a pilot error. However, the FAA has delegated to the PIC the responsibility of determining the airworthiness of the aircraft. This includes the proper paperwork. The PIC must be familiar with all the aircraft's paperwork and thoroughly check each item before accepting an aircraft for flight, especially if recent maintenance has been performed. Maintenance items not written off could be maintenance not performed; therefore, the flight could be unsafe as well as illegal.

In the category of flight crew not checking with maintenance when a discrepancy exists (item 3 above), these 19 ASRS reports reveal classic human factors issues. As presented in Chap. 8, reasons for not checking with maintenance when a discrepancy existed are as follows: (1) There was no resident maintenance at airport. (2) Checking with maintenance would result in a departure delay. (3) Captain accepted the risk, sometimes against the objections of the first and second officers. (4) PIC did not know that maintenance was required. PIC did not check logbook.

The psychological factors involved are (1) judgment and decision making and (2) failure to use accepted procedures, in other words, a lack of self-discipline.

Item 4 above is the disagreement between flight crew and maintenance and/or management. As discussed in Chap. 9, attempting to satisfy the demands of management and the airline's maintenance department as well as save his own career and ensure flight safety sometimes puts the PIC in a compromising position, as indicated in these 15 ASRS reports.

Although each of these cases could be considered pilot error, it is hard to fault the PIC who is confronted with such a formidable adversary as maintenance backed by management. Possibly, crew resource management programs should include pilots and maintenance personnel, as well as members of management.

There are all sorts of reasons for lack of self-discipline. There are also many organizational situations where

Summary and Final Discussion 215

supervisory imposition of discipline is not easy or effective. There is only one solution to either case. *That solution is integrity*—a quality within that maintains the state of order and submission to rules and authority without the need for anyone besides the pilot him- or herself.

Two contributing factors are seen in all the ASRS reports:

1. Pressure for on-time departures
2. Pilots not knowledgeable enough regarding the aircraft's systems

We will discuss item 2 above first. Yes, the airline provides comprehensive training programs on the aircraft's systems. However, as mentioned previously, possibly the pilot may be too dependent on them, expecting the training program to cover all eventualities.

For example, in ASRS report #154209 (Chap. 7), the reporter stated that he had no specific training regarding the procedure to manually remove the gear uplock pin since it was included in the hydraulic system emergency procedures, not the "partial gear, or gear-up landing checklist." Possibly, a more thorough knowledge of all systems and their interfaces would alleviate many problems, especially in flight, and assist in the decision-making process.

Since pilots are type-rated for a particular aircraft (B-727, for example), they usually fly the same aircraft type for years. As discussed in Chap. 3, acquiring intimate knowledge of the aircraft's systems, even through self-study, should not be too formidable an undertaking. Also, consulting with mechanics during slack time is a good way learn more about various systems. The mechanics' system training programs are more comprehensive than those presented to pilots.

Now we discuss item 1—pressure for on-time performance. As discussed in Chap. 3, by thoroughly familiarizing themselves with the aircrafts' systems, the pilots can intelligently discuss and even question maintenance on various procedures. Thus, they do not have to rely entirely on accepting maintenance's assessment of the problem. The pilot is in a more favorable position for making a decision regarding the airworthiness of the aircraft.

Summary

1. A pilot must be thoroughly familiar with the aircraft's systems.
2. Checklists and preflight procedures must be followed religiously.
3. Self-discipline is essential. Follow established rules and procedures.

APPENDIX

Typical Turbine-Powered Airline Aircraft and Its Systems

The majority of ASRS pilot reporters flew turbine-powered FAR part 121 air carrier aircraft. These are complex, aerodynamic, and mechanical airplanes. It is the responsibility of the airline maintenance department to keep these complex airplanes in airworthy condition and safe for flight. According to the FAA, it is the responsibility of the pilot-in-command to determine whether an aircraft is in safe condition (airworthy) to accept for flight.

The Aircraft

A typical medium-large transport, as operated by the world's leading airlines, is a low-wing, twin-engine, 100- to 300-passenger airplane with retractable landing gear. See FIG. A-1. Cruising speed is $M = 0.82$ to 0.85, or about 550 mph at altitude. The European Airbus A300, as operated by many U.S. airlines, is described below as a typical example.

220 Appendix

A-1 *The Airbus A300 twin-jet airliner. (Courtesy Airbus Industrie.)*

Design Features

Aerodynamics

Airplanes that fly at speeds close to Mach 1 employ wings of about 35° sweepback. Sweepback delays the formation of shock waves, allowing the wing to fly at higher Mach numbers than a straight wing, without producing compressibility drag.

Airfoil sections are usually modified versions of the NASA-developed supercritical section. This airfoil section generates significantly more lift over the rear portion of the wing than do more conventional airfoils. See FIG. A-2.

Since the basic wing is designed for high speeds, the wing must be able to be modified in flight to accommodate lower speeds during takeoff, climb, approach, and

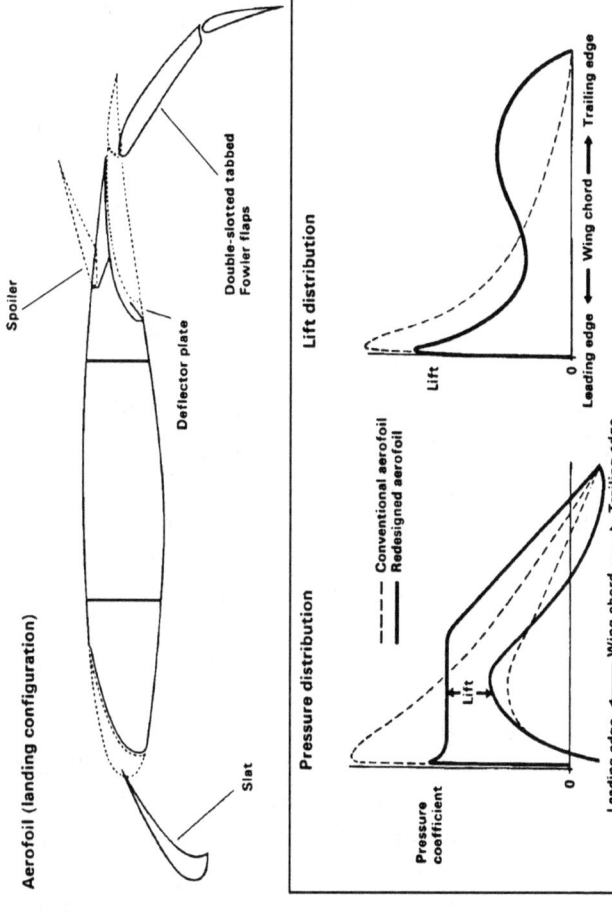

A-2 *Typical airfoil section (landing configuration) and pressure/lift characteristics. (Courtesy Airbus Industrie.)*

landing. FIGURE A-2 shows a typical airfoil with leading- and trailing-edge high-lift devices.

Flight controls

All flight controls are hydraulically operated with no manual reversion. Mechanical artificial feel systems (spring and dynamic pressure) are located near the control surfaces. See FIG. A-3.

Control surfaces are generally designed so that they accomplish the following:

- Safe flight is possible even after a control jam.
- No single fault, other than a jam, will cause a reduction of control capability.
- No combination of more than one failure will result in loss of control.
- No single pilot action can depressurize all three hydraulic systems simultaneously.

The low-speed aileron is locked at speeds of about 190 knots (kn). Pitch control is provided by conventional elevators with an adjustable horizontal stabilizer (tailplane) for trim. The tailplane is powered by three motors driving a fail-safe screwjack.

High lift is provided by tabbed Fowler flaps. These produce about a 25 percent increase in wing chord when fully extended. Lift is further augmented by use of leading-edge slats. The drive mechanisms for both slats and flaps are similar, each powered by twin motors driving ball screwjacks on each surface with built-in protection against asymmetric operation.

A lift dump capability providing positional ground adhesion is achieved by the combination of spoilers, air brakes, and two additional surfaces above the inboard flap.

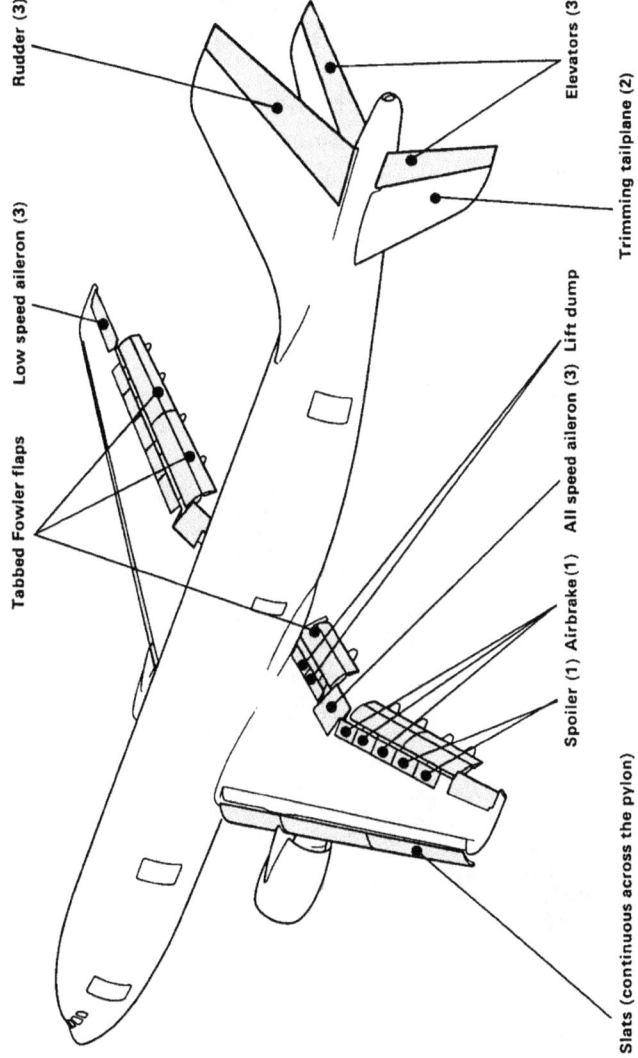

A-3 *Typical jetliner flight controls. (Number of servos per surface are indicated. The lift dump has one jack per surface, plus one to operate the spoilers for lift dump.)*

Structure

The basic structural material is high-strength aluminum alloy with steel or titanium for some major components. Nonmetallic materials (composites) are used in some areas of secondary structure.

The wing is integral with the fuselage—spars and skins continued across the center section. The wings' integral fuel tanks are sealed with a fuel-safe compound.

Power plant

The underwing engines are of the fuel-efficient, twin-spool fan-jet type, producing 50,000 lb of thrust each. FIGURE A-4 shows a Pratt & Whitney PW4000 fan-jet engine while FIG. A-5 illustrates basic turbofan engine instrumentation and controls.

A-4 *Pratt & Whitney's PW4000 commercial fan-jet engine. (Courtesy Pratt & Whitney.)*

Typical Turbine-Powered Aircraft 225

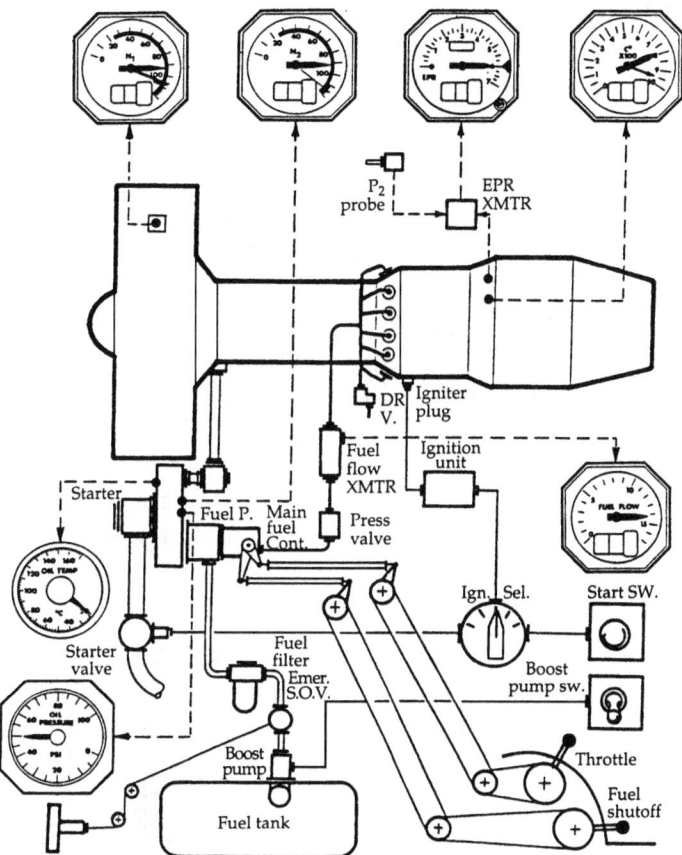

A-5 *Basic turbofan instrumentation and controls. (Courtesy General Electric.)*

Thrust reversers are incorporated into the fan-jet stream with an additional mechanism to reverse the hot jet thrust. Quick attach-detach fittings and couplings between pod and pylon enable the power plant to be removed and installed on the aircraft within about 2 hours under normal conditions.

Auxiliary power unit (APU)

The auxiliary power unit is installed in the tail cone of the fuselage, exhausting upward away from ground personnel. See FIG. A-6. The fire protection system is self-contained, and firewall panels protect the main structure from a fire hazard. Operation can be on the ground, in flight up to 30,000 feet (9100 m), and in icing conditions. Relights are possible up to 25,000 ft (7600 m).

The APU has a twin-spool compressor with interspool air bleed and shaft power extraction from the high-pressure spool, which is automatically controlled to a constant speed. See FIG. A-7.

The APU improves dispatchability and makes the aircraft completely independent of external power sources. The APU can be used for the following purposes in conjunction with the aircraft systems:

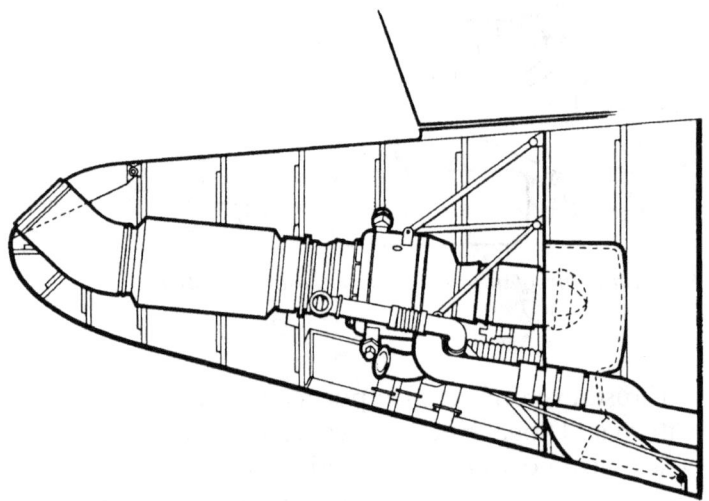

A-6 *The tail-mounted APU is an alternative source for in-flight systems power. (Courtesy Airbus Industrie.)*

Typical Turbine-Powered Aircraft 227

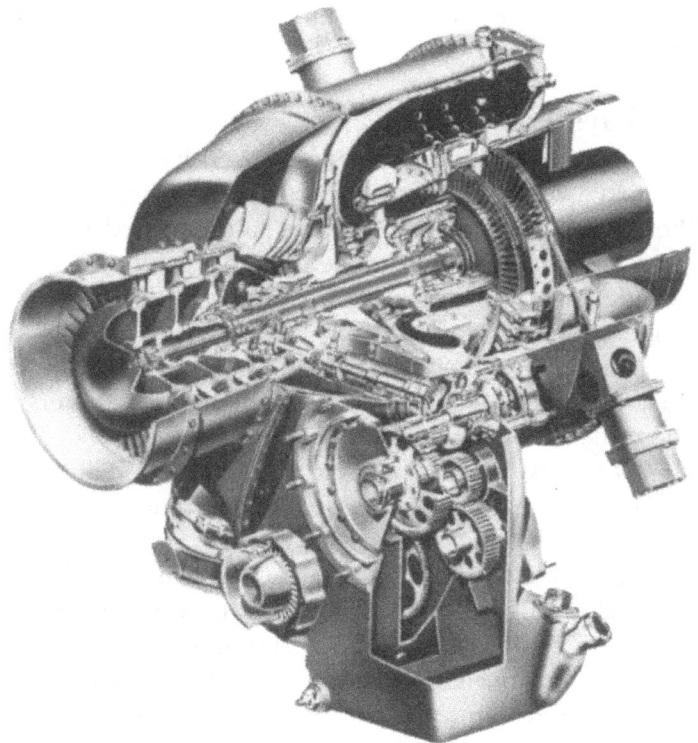

A-7 *A Garrett auxiliary power unit. (Courtesy Airbus Industrie.)*

- Engine starting (bleed air)
- Electric power—90 kVA nominal rating up to 30,000 ft
- Hydraulic power via two electric pumps in the green circuit
- Cabin air supply, both on the ground and in flight

APU starting is fully automatic, with power being supplied by the main batteries. It may also be started from the main engine generator or from a ground source.

Audio and visual fire warnings are provided for both flight and ground crews.

Fuel system

The four integral wing tanks (FIG. A-8) have combined capacity of 9460 lmp gal (11,360 U.S. gal, 43,000 L) on the Airbus aircraft. A vent surge tank is fitted in each wing tip. An integral fuel tank in the wing-center section for added range is included in the basic design of the longer-range variant.

Eight identical fuel pumps, two in each tank driven by flooded three-phase AC motors, are housed in canisters to allow their removal without draining the fuel tanks. The pair of pumps in each outer tank are located in a collector box, which is maintained full by a jet pump operated from associated fuel pumps. See FIG. A-9.

Normally the fuel system in each wing feeds the associated engine through a manually operated valve (which is replaceable without entering the fuel tank). Fuel in the inner tanks is consumed first. Changeover to the outer tanks is automatic, needing no crew action. In an emergency, suction feed is possible. The total fuel load is available to either engine through a cross-feed valve in the event of an engine shutdown. Two couplings on the underside of each wing allow refueling from either side in less than 20 minutes, at a total rate up to 500 Imp gal/min (2270 L/min).

The APU can be fed from either inner tank by using the respective main engine fuel pumps.

The main fuel gauging system is of the capacitance-bridge type and provides flight deck indication of fuel quantity and drives the fuel load preselector. The measuring units are positioned to keep aircraft attitude errors minimal. Magnetic level indicators provide a cross-check.

There is structural provision for a fuel jettison facility.

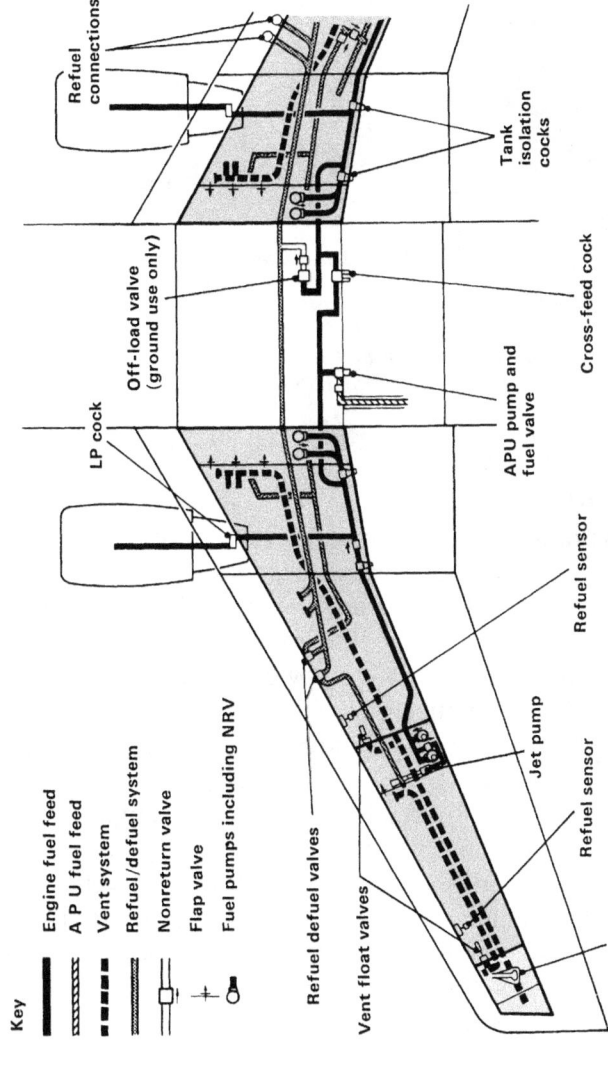

A-8 *A typical twin jetliner fuel system. (Courtesy Airbus Industrie.)*

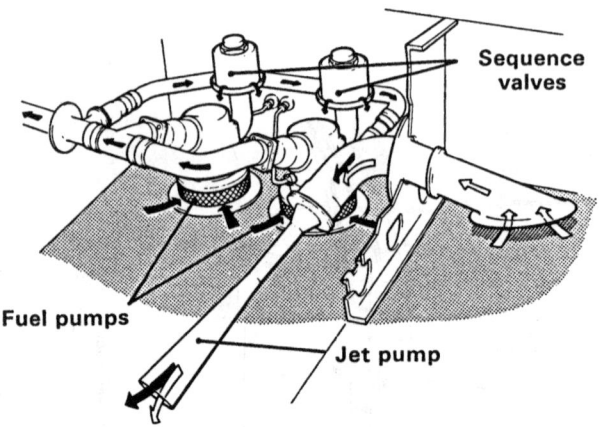

A-9 *The fuel pump compartment of the outer tank shows the sequence valves and jet pump. (Courtesy Airbus Industrie.)*

Hydraulic system

The hydraulic system comprises three completely independent circuits operating simultaneously. See FIG. A-10. The fluid used is a fire-resistant phosphate ester type, working at a pressure of 3000 lb/in^2.

The three circuits provide power for the primary flying controls. If any circuit should fail, the fault is indicated on the flight deck, but full control of the aircraft is retained without any crew action being necessary.

Each circuit is normally powered by engine-driven self-regulating pumps—one on each engine for the green circuit and one each for the blue and yellow circuits. An accumulator in each circuit compensates for the response time of the pumps and damps out transient pressure surges. If power is lost in the blue or yellow circuit, it can be restored through motor/pump transfer units driven by the green system, without any fluid changing circuit. Also, two parallel electrical pumps in

Typical Turbine-Powered Aircraft 231

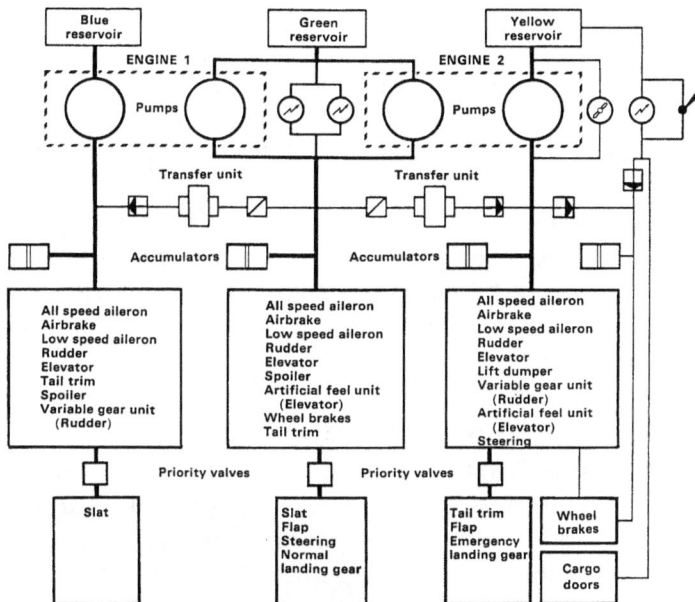

A-10 *The hydraulic system comprises three completely independent circuits operating simultaneously. (Courtesy Airbus Industrie.)*

the green circuit provide either in-flight emergency power or, by utilizing the transfer units, power to enable ground checks to be performed on the whole hydraulic system.

In the event of a double engine failure, power can be restored in the yellow circuit by a dropout ram air turbine driving a hydraulic pump.

If yellow circuit pressure is lost, the emergency wheel braking system is completely isolated by a nonreturn valve, and pressure is provided by a separate accumulator. A third electric pump recharges this accumulator for towing and parking and also operates the cargo doors. A hand pump provides additional means of operating the cargo doors.

Services not associated with flying can be isolated from the hydraulic supply either, like the landing gear, by auto-isolation on retraction or, like the flaps and slats, by a priority valve which closes if, because of fluid leakage, the pressure differential across it becomes too great.

The hydraulic power equipment is installed in the main landing gear wells with guards to protect it from damage. The equipment is grouped in modules so that it can be removed without interfering with piping or associated equipment.

Environmental control system

Air conditioning Air is supplied from the engine compressors, the APU, or a high-pressure ground source and is controlled by separate and parallel units, each of which includes a flow-limiting unit, heat exchanger, cold air unit, water separator, and a temperature control and precooler unit. See FIG. A-11.

Conditioned air is distributed to four zones in the pressure cabin, one of which is the flight deck. Temperature in each zone can be independently controlled between 18 and 29°C (65 and 85°F).

Pressurization Two independent automatic systems, with manual override, control cabin altitude, its rate of change, and the differential pressure. Only one system operates at any one time, the second being standby.

The cabin differential pressure for normal operation is 8.25 lb/in^2 (570 mb); i.e., sea-level cabin conditions can be maintained up to 21,000 ft (6400 m), while at 35,000 ft (10,600 m) the cabin altitude is about 6000 ft (1800 m). The rate of change of cabin altitude can be controlled between 200 and 2000 ft/min (1 and 10 m/s).

Typical Turbine-Powered Aircraft

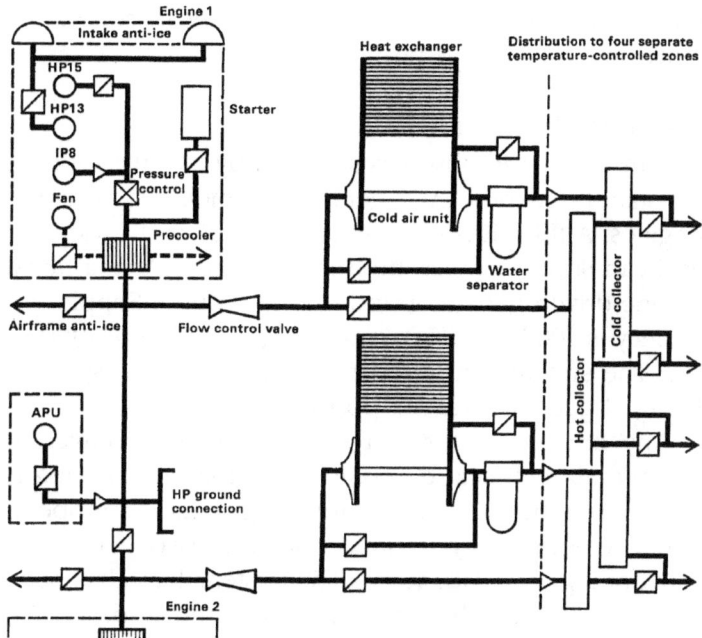

A-11 *Cabin air conditioning is provided by two separate and parallel units supplied by engine compressor bleed air, APU, or a high-pressure ground air source. (Courtesy Airbus Industrie.)*

Safety Valves limit the maximum differential pressure to 9.25 lb/in^2 (638 mb) should the pressure control and air conditioning system fail.

Ice protection Hot air protection is provided for the engine intakes and the slat sections on the wings outboard of the engines. An electrical heating system ensures that the windshields are kept free from ice and mist.

Fault detection In addition to normal flight deck indicators and warning lights, fault detection systems are installed to allow main failures to be diagnosed, in particular temperature controls, electrical continuity, electronic boxes, and probes.

Electrical system

The main electric power is supplied by two three-phase constant-frequency AC generators mounted on the engines. A third identical generator is driven by the APU and can supply power, both in flight to replace a failed engine-driven generator and on the ground. Supply frequency is 400 hertz (Hz), and voltage is 115/200 volts (V) AC. The system comprises two separate channels; paralleling of generators is not possible.

Any one generator can supply sufficient power to operate all technical equipment. No load monitoring is necessary when an engine-driven generator is inoperative. Takeoffs are permitted with any two generators functioning. See FIG. A-12. The system has also been designed to meet the requirements of autoland.

AC generators Each engine-mounted generator is driven at 8000 rpm by a constant-speed drive (CSD) unit. The generator and CSD are separately mounted on opposite sides of the engine gearbox.

Since the APU is automatically controlled to very close tolerances, its generator is driven at constant speed through its gearbox. The brushless air-cooled generators each have a nominal and continuous rating of 90 kVA at the busbar; overload ratings allowed are 135 kVA for 5 minutes, 180 kVA for 5 seconds.

DC power Three unregulated silicon transformer-rectifier units (TRUs) supply 28-V DC power and are cooled by natural ventilation, which is fan-assisted if half load is exceeded. Normal DC power is supplied by two TRUs with the third retained for essential services. No DC load shedding is required after the loss of one TRU. A fourth identical TRU converts ground AC power to DC.

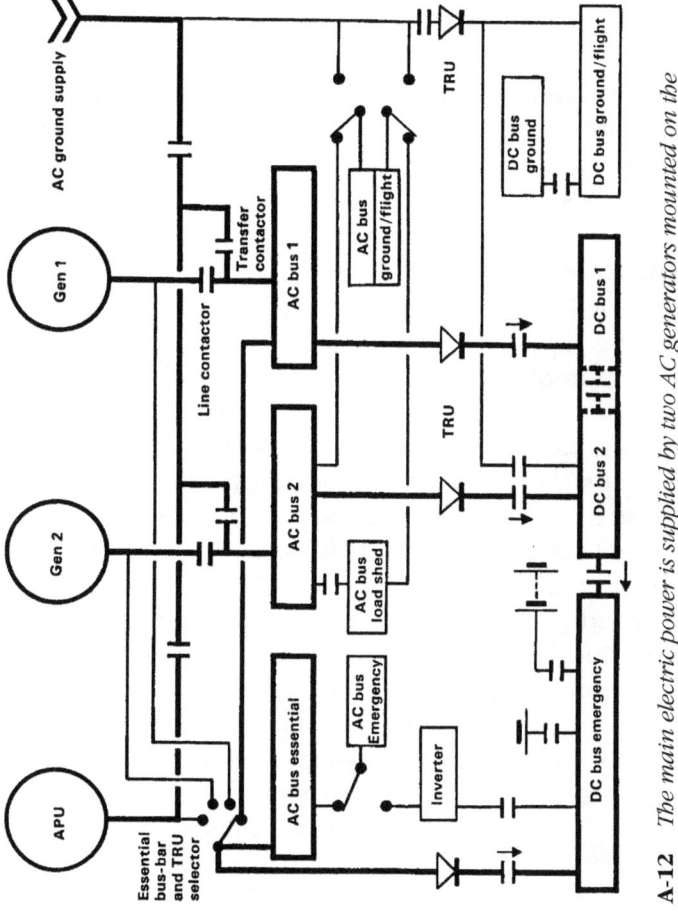

A-12 *The main electric power is supplied by two AC generators mounted on the engines. A third AC generator is driven by the APU. (Courtesy Airbus Industrie.)*

Emergency power Aircraft batteries supply direct current and, through the 115-V 400-Hz inverter, single-phase AC power for the following:

- APU starting and fuel control
- Standby lights
- Emergency power

The emergency system is completely isolated from the main system with sufficient power available, after three attempts at an APU start, to continue flight for a further 25 minutes.

Flight deck or cockpit

FIGURE A-13 illustrates a typical instrument panel layout with the captain's flight instruments on the left and first officer's duplicate flight instruments on the right. The center panel displays the engine instruments, master warning panel, standby flight instruments, landing gear and flaps/slats lever, and flight controls indicator.

Systems monitoring is done from a control panel on the right-hand side of the flight deck. The segregation of controls and instruments between this station and the overhead panel is such that those required for immediate action during flight are within easy reach of the three crew members while facing forward.

Most FAR part 121 air carrier aircraft are certified for category 3 automatic landing (autoland) capability. However, not all airports and/or aircraft are certified for category 3. FIGURE A-14 shows the different low-visibility landing criteria.

Ground handling/servicing

Servicing points and equipment bays are located to ensure quick and easy accessibility and maximum utilization of servicing equipment. Utilization of vehicles as

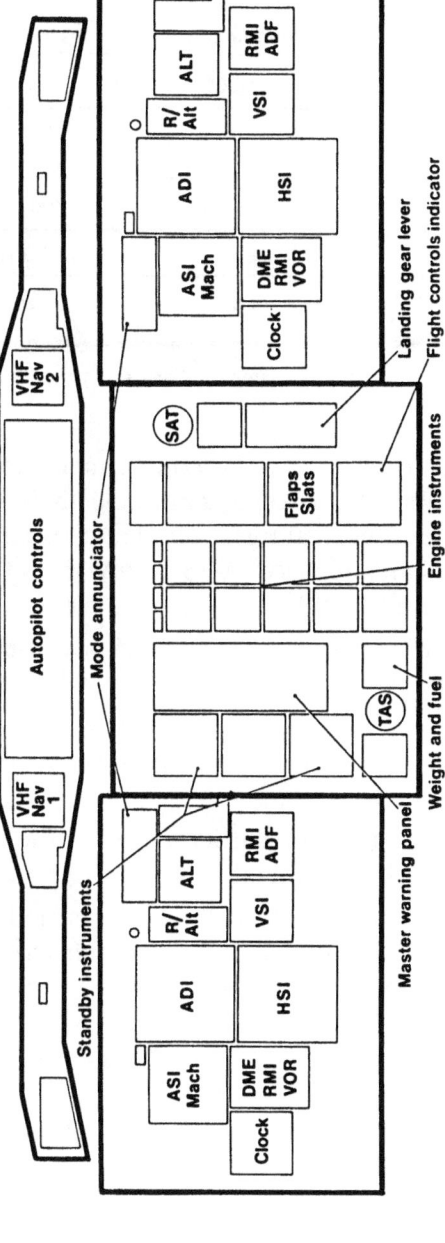

A-13 *Typical flight deck instrument panel layout. (Courtesy Airbus Industrie.)*

	Decision height ft	Visibility ft (m)
Cateogry 1	above 200	2,600 (800)
2	200–100	2,600–1,200 (800–400)
3 A	below 100	700 (200)
3 B	zero	150 (50)
3 C	zero	zero

A-14 *Low-visibility landing criteria. (Courtesy Airbus Industrie.)*

servicing platforms is a major consideration in locating servicing points. Turnaround time objectives are set at 30 minutes for a normal turnaround and 20 minutes for a fast turnaround. See FIGS. A-15 and A-16.

To achieve these objectives:

- The APU can provide all ground power needs, reducing the number of servicing vehicles and ground personnel.
- Jetways can be brought up to all or any of the six cabin doors.
- All servicing can be carried out from the right-hand side of the aircraft.
- Baggage and freight loading, also carried out from the right-hand side, is largely containerized and power-operated.

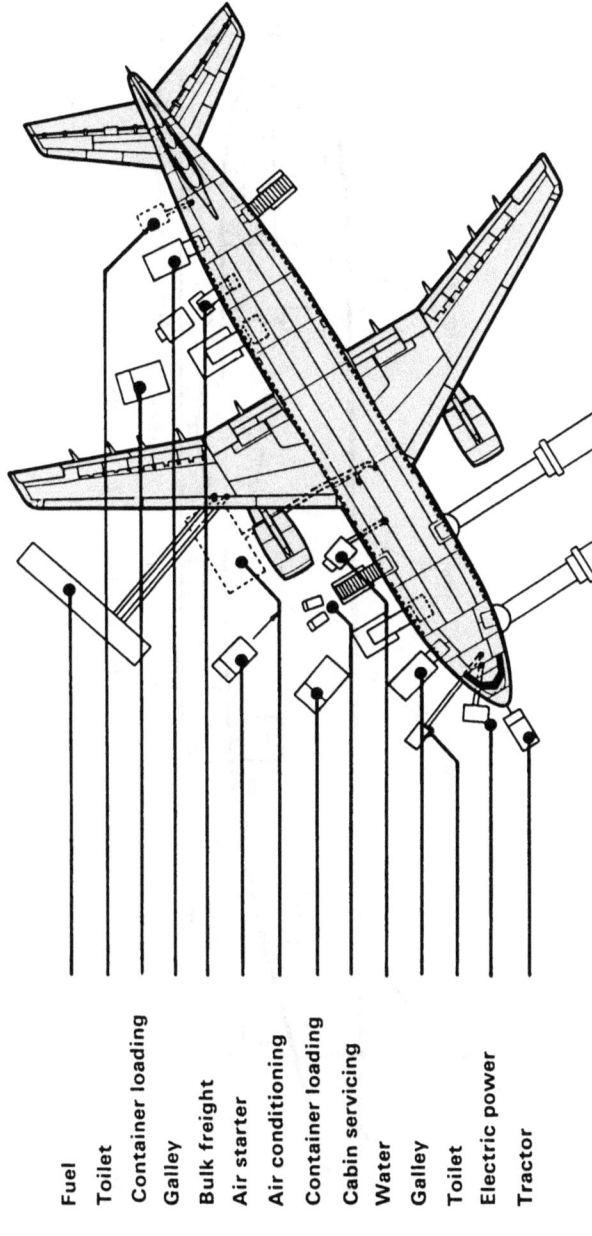

A-15 Maximum servicing vehicle requirement—APU not running. (Courtesy Airbus Industrie.)

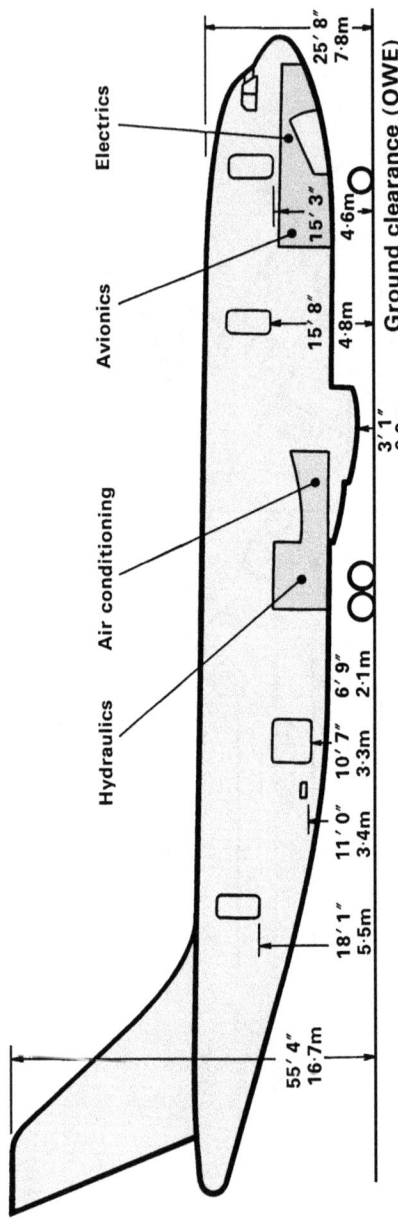

A-16 *Servicing points and equipment bays located for easy accessibility. (Courtesy Airbus Industrie.)*

Bibliography

Adamski, A. J., and T. J. Doyle. *Aviation Regulatory Process*. Hayden-McNeil Publishing, Inc., Westland, Mich., 1995.

Krause, Shari Stamford. *Aircraft Safety*. McGraw Hill, New York, 1996.

Spence, Charles F. *AIM/FAR 2000*. McGraw Hill, New York, 1999.

Wood, Richard H. *Aviation Safety Programs: A Management Handbook*. Jeppesen Sanderson, Inc., Englewood, Colo., 1991.

Index

A

abnormal procedures, 123–134
accidents, 3, 4
aerodynamics, 220
air carrier:
  certificate, 9
  freight (see ASRS reports, type of operation)
  major airlines, FAR part 121, 9, 219
  operations (see ASRS reports, type of operation)
air conditioning and pressurization, 73, 189, 232
air taxi:
  FAR part 135, 10
  operations (see ASRS reports, type of operation)
air traffic control operator, 10
Air Transport Association (ATA), 13
aircraft operating certificate:
  FAR part 119, commuter operations, 10
  FAR part 121, air carrier, 9
  FAR part 125, business aircraft operation, 10
  FAR part 135, air taxi and some commuter operations, 10
Aircraft Owners and Pilots Association (AOPA), 13
aircraft systems:
  description, 217–240
  flight crew misunderstanding of, 115–119
aircraft type rating, 7, 215
aircraft types (see ASRS reports, aircraft involved)
Airline Pilots Association (ALPA), 13, 35
airplane flight manual, 12, 14
airworthiness:
  certificate, 12, 67, 78, 79, 92
  determination of, 11, 50, 53, 61, 177, 206

airworthiness (Cont.):
  inspector (FAA), 9
  responsibility for, 12
  standards, 8
approval for service, 9
APU (see auxiliary power unit)
ASRS incident reports:
  abnormal procedures, 123–134
  checklist, missed items, 95–112
  disagreement, flight crew and maintenance and/or management, 177–208
  emergency procedures, 123–134
  maintenance, flight crew not checking with, 137–174
  MEL, problems with, 17–50
  paperwork, improper maintenance, 65–93
  preflight, 95–112
  systems, misunderstanding of, 113–119
ASRS reports, aircraft involved:
  reciprocating engine powered:
    small, less than 5000 lb, 71, 77, 78, 143, 155, 190
    small transport, 5001–14,500 lb, 88
  specific aircraft:
    Airbus Industrie A-300, 220
    Beech 1900, 106
    Boeing 737-200, 21, 130
    Brasilia EMB-120, 87
    Canadair CL39, 65, 170
    Cessna 402 C, 90
    deHaviland Dash 8, 110
    Falcon DA 20, 206
    Fokker FK 10, 165
    McDonnell Douglas DC8-73F, 171
    McDonnell Douglas MD 90, 164
    McDonnell Douglas super 80, 46
  turbojet powered:
    large, over 300,000 lb, 30, 54, 100, 124, 148, 158, 200, 202

243

244 Index

ASRS reports, aircraft involved, turbojet powered (*Cont.*):
  light, 14,501–30,000 lb, 74
  medium, 30,001–60,000 lb, 37, 82, 170, 205
  medium large, 60,001–150,000 lb, 20–24, 26–33, 37, 39, 43–48, 55, 58, 68, 69, 73, 80, 83, 99, 115–118, 126–132, 137–143, 147, 153, 156, 164–166, 192
turboprop powered:
  light, 14501–30,000 lb, 24, 36, 103, 106, 145, 151, 167, 177, 179, 184, 187, 189
  medium, 30,001–60,000 lb, 37, 46, 85, 87, 109, 138, 156, 160, 196
  small, 5001–14,500 lb, 56, 105
ASRS reports, type of operation:
  air carrier, commuter passenger flight, 24, 36–39, 46, 56, 145, 151, 156, 160, 166–171, 179, 184–190
  air carrier, freight, 41, 54, 137, 171, 204
  air carrier (major airlines) passenger flight, 20–24, 26–35, 39, 42–46, 55, 58, 68–70, 72, 80–88, 99–102, 109, 115–117, 123–132, 139–143, 146–151, 153, 158, 162–166, 177, 181, 192–204
  air taxi, 88–90, 143
  corporate flight, 74, 103, 105
  general aviation/training, 154, 190
ATIS (*see* automatic terminal information service)
autoland (automatic landing), 236
automatic terminal information service (ATIS), 145
autopilot, 26, 105
auxiliary power unit (APU), 143, 145, 226–227
aviation safety reporting system (ASRS), 4

C

captain (pilot-in-command), 13
CDL (*see* configuration deviation list)
certification:
  of aircraft, 3, 6
  of dispatcher, 10

certification (*Cont.*):
  of mechanic, 10
  of pilot, 7
checklist, 14, 97–112, 123
cockpit voice recorder, 22
commuter air carrier:
  FAR 135 and 119, 10
  operations (*see* ASRS reports, type of operation)
configuration deviation list (CDL), 22, 195
constant speed drive, 2
contract maintenance (mechanic), 21, 42
copilot (*see* first officer)
corporate flight operations (*see* ASRS reports, type of operation)
crew complement, air carrier aircraft, 13
crew coordination, 133
crew resource management (CRM), 15, 142, 151, 212, 214
CRM (*see* crew resource management)

D

disagreement, flight crew and maintenance and/or management, 177–208
disciplinary action, 5
dispatcher, 10, 11, 44, 72
door latch problems, 81, 197
dripstick, 24, 47

E

EGT (*see* exhaust gas temperatures)
electric trim inoperative, 184–187
electrical system:
  description, 234–236
  problems, 86, 141, 151
emergency:
  cockpit checklist (*see* checklist)
  lights, 36
  procedures, 123–134
engine:
  flameout, 103
  ignition systems problems, 116, 147
  overtemperature (EGT), 115, 162, 202–204
  shutdown in flight, 137–139
  start fire, 192–194
environmental control system, 232, 233

Index 245

exhaust gas temperature (EGT), 15, 162, 202–204

F

F/A (*see* flight attendant)
FAA (*see* Federal Aviation Administration)
FAA inspector, 104, 144
FAR (*see* Federal Aviation Regulations)
F/E (*see* flight engineer)
Federal Aviation Administration (FAA), 4
Federal Aviation Regulations (FAR):
  aircraft type rating, 7
  maintenance highlights, 6
  part 23, 8
  part 25, 8
  part 43, 8
  part 61, 7
  part 65, 10
  part 91, 11
  part 119, 10
  part 121, 9
  part 125, 10
  part 135, 10
ferry flight (permit) 14, 72, 106–109
fire warning, 156
first officer (F/O), 7, 13
fixed-base operator (FBO), 67
flameout, 103
flight attendant (F/A), 13, 45
flight controls, 222–223
flight deck (cockpit), 236–237
flight engineer (F/E), 101
flight restrictions, 38, 44, 107, 213
F/O (*see* first officer)
freighter aircraft (*see* ASRS reports, type of operation)
fuel system:
  description, 228–230
  dripstick, use of, 24, 47
  exhaustion of fuel, 123–125
  imbalance, 123–125
  inoperative gauge, 23, 46
  leakage, 156
  transfer, 153

G

gear (landing) problems:
  door positioned improperly, 39, 58
  door removed, 37
  extension problems, 166

gear (landing) problems (*Cont.*):
  ferry flight, gear down, 106
  landing, one gear up, 126
  would not retract, 100
general aviation, 6, 13, 15, 67, 68, 76, 78, 91, 154, 190
  (*See also* ASRS reports, type of operation)
general operating rules, FAR part 91, 11
ground handling/servicing, 236, 238–240

H

hail damage, 84, 204–206
human error, human factor, 3, 5, 111, 124, 214
hydraulic system, 123, 129, 130, 215, 222, 230–232

I

IMC (*see* instrument meteorological conditions)
immunity, 5, 6, 75
incident reporting system, 4
  (*See also* aviation safety reporting system)
incident reports (*see* ASRS incident reports)
inoperable instruments and equipment, 12, 19
instrument meteorological conditions (IMC), 4, 167
instrument panel, 236, 237
integrity, 112, 174
intimidation, 55

L

landing gear [*see* gear (landing) problems]
landing mode category, 28, 238
logbook, 9, 14, 28, 76, 80, 83, 85, 145, 151, 194, 200

M

maintenance:
  contract, 21, 42, 67
  deferred, 56
  definition, FAR part 43, 8, 12
  discrepancy, 21, 30, 36
  inspector, FAA, 27, 75
  paperwork, 67–93
  preventive, 8, 12

mechanic, A&P, 9–11, 72, 191
minimum equipment list (MEL), 12, 14, 19–49, 121, 186, 188, 197, 213, 185, 194, 197, 213

N
NASA (*see* National Aeronautics and Space Administration)
National Aeronautics and Space Administration (NASA), 5
National Aviation System (NAS), 5
National Business Aircraft Association (NBA), 13
National Transportation Safety Board (NTSB), 4
NTSB (*see* National Transportation Safety Board)

O
on-time departure, excessive pressure for, 53–63, 215
operating certificate, 9
operation, types of (*see* ASRS reports, type of operation)
oxygen service door, 169, 172

P
pack (*see* air conditioning and pressurization; environmental control system)
PIC (*see* pilot-in-command)
pilot, certification of, 7
pilot error, 53, 212
pilot-in-command (PIC), 13, 50, 53, 61
  responsibilities, 11, 67, 72, 117, 206, 212
power settings, turbojet engine, 15
preflight, 97–112
pressure for on-time departures, 49, 53–63, 111, 215, 216
pressurization, 73, 82, 188, 197, 232
preventative maintenance (*see* maintenance)

R
reciprocating engine powered aircraft (*see* ASRS reports, aircraft involved)
registration certificate, 12, 67, 74, 78, 79, 92
responsibility of PIC (*see* pilot-in-command)
reverser (*see* thrust reverser)

S
safety, 6, 211, 212
scheduled maintenance (*see* maintenance)
second officer (S/O), 13
self-discipline, 111, 174
skin damage, 139–140
S/O (*see* second officer)
stabilizer problems, 198–200
structure, 224
systems, turbine powered aircraft, 219–240

T
tail damage, 99
technical standard order (TSO), 192
thrust reverser, 158, 225
training, 15, 62, 123, 211, 212, 215
TSO (*see* technical standard order)
turbojet powered aircraft (*see* ASRS reports, aircraft involved)
turboprop aircraft (*see* ASRS reports, aircraft involved)
type rating (*see* aircraft type rating)

U
uncommanded rudder displacement, 130–133

W
warning horn, takeoff, 59
warning panel, 164, 165
  fault display unit, 165–166
weight and balance, 11, 26, 74
wingtip light, 41

www.ingramcontent.com/pod-product-compliance
Lightning Source LLC
Chambersburg PA
CBHW060114170426
43198CB00010B/890